Quantitative Methods in the Humanities and Social Sciences

Editorial Board
Thomas DeFanti, Anthony Grafton, Thomas E. Levy, Lev Manovich, Alyn Rockwood

Quantitative Methods in the Humanities and Social Sciences is a book series designed to foster research-based conversation with all parts of the university campus – from buildings of ivy-covered stone to technologically savvy walls of glass. Scholarship from international researchers and the esteemed editorial board represents the far-reaching applications of computational analysis, statistical models, computer-based programs, and other quantitative methods. Methods are integrated in a dialogue that is sensitive to the broader context of humanistic study and social science research. Scholars, including among others historians, archaeologists, new media specialists, classicists and linguists, promote this interdisciplinary approach. These texts teach new methodological approaches for contemporary research. Each volume exposes readers to a particular research method. Researchers and students then benefit from exposure to subtleties of the larger project or corpus of work in which the quantitative methods come to fruition.

More information about this series at http://www.springer.com/series/11748

Dirk Speelman • Kris Heylen • Dirk Geeraerts
Editors

Mixed-Effects Regression Models in Linguistics

Editors
Dirk Speelman
Faculty of Arts, Research Group QLVL
KU Leuven, Belgium

Kris Heylen
Faculty of Arts, Research Group QLVL
KU Leuven, Belgium

Dirk Geeraerts
Faculty of Arts, Research Group QLVL
KU Leuven, Belgium

ISSN 2199-0956 ISSN 2199-0964 (electronic)
Quantitative Methods in the Humanities and Social Sciences
ISBN 978-3-319-69828-1 ISBN 978-3-319-69830-4 (eBook)
https://doi.org/10.1007/978-3-319-69830-4

Library of Congress Control Number: 2018930011

© Springer International Publishing AG 2018
This work is subject to copyright. All rights are reserved by the Publisher, whether the whole or part of the material is concerned, specifically the rights of translation, reprinting, reuse of illustrations, recitation, broadcasting, reproduction on microfilms or in any other physical way, and transmission or information storage and retrieval, electronic adaptation, computer software, or by similar or dissimilar methodology now known or hereafter developed.
The use of general descriptive names, registered names, trademarks, service marks, etc. in this publication does not imply, even in the absence of a specific statement, that such names are exempt from the relevant protective laws and regulations and therefore free for general use.
The publisher, the authors and the editors are safe to assume that the advice and information in this book are believed to be true and accurate at the date of publication. Neither the publisher nor the authors or the editors give a warranty, express or implied, with respect to the material contained herein or for any errors or omissions that may have been made. The publisher remains neutral with regard to jurisdictional claims in published maps and institutional affiliations.

Printed on acid-free paper

This Springer imprint is published by the registered company Springer International Publishing AG part of Springer Nature.
The registered company address is: Gewerbestrasse 11, 6330 Cham, Switzerland

Preface

When data consist of grouped observations or clusters, and there is a risk that measurements within the same group are not independent, group-specific random effects can be added to a regression model in order to account for such within-group associations. Regression models that contain such group-specific random effects are called mixed-effects regression models, or simply mixed models. Mixed models are a versatile tool that can handle both balanced and unbalanced datasets and that can also be applied when several layers of grouping are present in the data; these layers can either be nested or crossed.

In linguistics, as in many other fields, the use of mixed models has gained ground rapidly over the last decade. This methodological evolution enables us to build more sophisticated and arguably more realistic models but, due to its technical complexity, also introduces new challenges. This volume brings together a number of promising new evolutions in the use of mixed models in linguistics but also addresses a number of common complications, misunderstandings, and pitfalls. Topics that are covered include the use of huge datasets, dealing with non-linear relations, issues of cross-validation, and issues of model selection and complex random structures. The volume features examples from various subfields in linguistics. The book also provides R code for a wide range of analyses.

The idea for this book first arose at the 2012 Leuven Statistics Days conference, the theme of which was 'Mixed models and modern multivariate methods in linguistics' (http://lstat.kuleuven.be/research/lsd/lsd2012/index.htm). The conference took place at the KU Leuven and was co-organized by LStat (Leuven Statistics Research Centre) and the linguistic research group QLVL. We thank all conference participants for their contributions to the conference. We also thank all authors for contributing to this book, and we thank all anonymous referees for their important criticisms.

Leuven, Belgium
January 2018

Dirk Speelman
Kris Heylen
Dirk Geeraerts

Contents

1 **Introduction** 1
  Dirk Speelman, Kris Heylen, and Dirk Geeraerts

2 **Mixed Models with Emphasis on Large Data Sets** 11
  Geert Verbeke, Geert Molenberghs, Steffen Fieuws, and Samuel Iddi

3 **The L2 Impact on Learning L3 Dutch: The L2 Distance Effect** 29
  Job Schepens, Frans van der Slik, and Roeland van Hout

4 **Autocorrelated Errors in Experimental Data in the Language Sciences: Some Solutions Offered by Generalized Additive Mixed Models**.. 49
  R. Harald Baayen, Jacolien van Rij, Cecile de Cat and Simon Wood

5 **Border Effects Among Catalan Dialects**................................... 71
  Martijn Wieling, Esteve Valls, R. Harald Baayen, and John Nerbonne

6 **Evaluating Logistic Mixed-Effects Models of Corpus-Linguistic Data in Light of Lexical Diffusion**.. 99
  Danielle Barth and Vsevolod Kapatsinski

7 **(Non)metonymic Expressions for GOVERNMENT in Chinese: A Mixed-Effects Logistic Regression Analysis**........................... 117
  Weiwei Zhang, Dirk Geeraerts, and Dirk Speelman

Chapter 1
Introduction

Dirk Speelman, Kris Heylen, and Dirk Geeraerts

Abstract As in many other fields, the use of mixed models has recently gained ground rapidly in linguistics. This methodological evolution enables us to build more sophisticated and arguably more realistic models, but, due to its technical complexity, also introduces new challenges. This volume brings together a number of promising new evolutions in the use of mixed models in linguistics, as well as addressing a number of common complications, misunderstandings, and pitfalls. Topics that are covered include the use of huge datasets, non-linear relations, issues of crossvalidation, and issues of model selection and complex random structures. The volume features examples from various linguistic subfields. This introductory chapter succinctly sketches how and why linguistic data often lend themselves to the use of mixed models and introduces the issues raised, and the topics covered, in the chapters of this volume.

1 Mixed Models

When data consist of grouped observations, and there is a risk that measurements within the same group are not independent, group-specific random effects can be added to a regression model in order to account for such within-group associations. Regression models that contain such group-specific random effects are called mixed-effects regression models, or simply mixed models. Mixed models are a versatile tool that can handle both balanced and unbalanced datasets and that can also be applied when several layers of grouping are present in the data; these layers can either be nested or crossed.

As in many other fields, the use of mixed models has recently gained ground rapidly in linguistics. This methodological evolution enables us to build more sophisticated and arguably more realistic models, but, due to its technical

---

D. Speelman (✉) · K. Heylen · D. Geeraerts
Faculty of Arts, Research Group QLVL, KU Leuven, Belgium
e-mail: dirk.speelman@kuleuven.be; kris.heylen@kuleuven.be; dirk.geeraerts@kuleuven.be

© Springer International Publishing AG 2018
D. Speelman et al. (eds.), *Mixed-Effects Regression Models in Linguistics*,
Quantitative Methods in the Humanities and Social Sciences,
https://doi.org/10.1007/978-3-319-69830-4_1

complexity, also introduces new challenges. This volume brings together a number of promising new evolutions in the use of mixed models in linguistics, as well as addressing a number of common complications, misunderstandings, and pitfalls. Topics that are covered include the use of huge datasets, non-linear relations, issues of cross-validation, and issues of model selection and complex random structures. The volume features examples from various linguistic subfields.

2 Mixed Models in Linguistics

Examples of early adoptions of mixed models in linguistics can be found in different subfields of linguistics. Some examples are [5] in corpus linguistics, [2, 6, 7] in psycholinguistics, [8] in sociolinguistics, [10] in dialectometry/dialectology, etc. One publication that deserves special mention is the 2008 textbook [1], which offers a comprehensive coverage of mixed models in linguistics and has been very instrumental in the wider adoption of this technique in linguistics.

Over the last decade, mixed models have become increasingly popular in linguistics. This has happened for good reasons, since several types of grouping are very common in linguistic data. The following paragraphs list but a few examples, and certainly do not exhaust all possible types of grouping in linguistic data.

- In corpus data, when we study some linguistic variable, sometimes several attestations of that variable were produced by the same *speaker/writer*. In that case, it is possible that instances produced by the same speaker/writer are not independent. Additionally, corpus data are often sampled from a mixture of *genres/text types*, and the utterances in the corpus touch upon different *topics*. Here, again, it is possible that observations within the same genre/text type or observations that share the same topic are not independent. Unfortunately, it is not always easy to 'tag' each observation for speaker, genre or topics information; this can be difficult or even impossible, either because metadata such as speaker information are missing or because it is hard to come up with a proper, or useful, classification of things such as genre or topic. A classification that is often used as a proxy for sources of grouping that are hard to detect directly, is that of the individual *texts/documents* in a corpus. The rationale is that for a number of reasons (which themselves are often hard to identify or disentangle), it can be the case that observations from the same text/document (e.g. the same conversation) are not independent, and that it therefore makes sense to treat individual texts/documents as grouping level. Also, there is the possibility that linguistic variables behave differently depending on their immediate linguistic context. For instance, this may depend on the specific *words* the variables occur with. For instance, a syntactic variation pattern (e.g. a word order alternation pattern) may behave differently, depending on what is the specific main verb in the pattern or in the syntactic context of the pattern.

- In experimental data, often several measurements apply to the same *participant* or to the same *linguistic stimulus*, or to the same combination of both (as is often the case in longitudinal studies).
- In survey data, the *location of residence* of an informant, or his/her *location of origin*, his/her *mother tongue*, etc. can also be possible sources of observation grouping. Also, when asked several similar questions, the *informant* himself/herself can be the source of non-independence of observations.
- In the context of language and education and language acquisition, *students/pupils*, *classes*, *schools*, *cities*, etc. can all be possible sources of observation grouping.

As these examples make clear, linguistic data often lend themselves to the use of mixed models. That being said, the application of mixed models to linguistic data often is far from a trivial matter, for a number of reasons. The following list contains some of these reasons. None of these are unique to linguistics, but together they sometimes make the application of mixed models to linguistics a complicated matter.

- It can be hard to distinguish between random-effect and fixed-effect factors. The distinction is clear in prototypical cases. In a prototypical fixed-effect factor, the variable has a rather limited set of levels, both in the sample and in the population, and all the levels that occur in the population also occur in the sample. In a replication study, the levels that occur in the new sample would be the same as in the original study. In a prototypical random-effect factor, the variable typically has a very large set of levels, certainly in the population, and the levels that occur in the sample are a random subset of the levels that occur in the population. In a replication study, the levels that occur in the new sample would typically differ (to a large extent, if not completely) from those in the original study. Treating such variables as random-effect factors allows us to build models the merits of which are not confined to the specific set of levels that were attested in our sample.

    Whereas variables such as *speaker/writer*, *participant*, *informant* or *word* in many studies approximate the prototypical case of random-effect factors, the situation is rather different for things such as *genre/text type* and *topic*. For those types of variables, two competing approaches both have their merits. One approach would be to opt for coarse-grained classifications (of genres/text type or topics) that can be used as fixed-effect factors. They would have to be such that all levels are attested in the sample, and that the levels jointly cover all situations we want to model. This may imply that we can have no ambition to extrapolate our findings beyond a somewhat restricted, but still rather broad, set of contexts (e.g. three or four broadly defined genres). The alternative approach is to work with much more fine-grained classifications (or even treat each individual text as a separate 'level'), which then are treated as random-effect factors. Of course, to some extent the choice between both approaches is related to our research goals. In some studies, difference between specific genres will be at the heart of the study. In other studies, genre could be considered a nuisance variable.
- Many variables in linguistics, including response variables, are categorical, especially in corpus linguistics. Therefore, the most often used type of mixed

models is that of mixed-effects logistic regression models. From a mathematical point of view, as far as modeling random effect structures is concerned, this is not the most favorable case. Moreover, the Zipfian distribution of word frequencies (with a few very high frequency words and many very low frequency words), as well as the typically very skewed distribution of the amount of utterances per speaker/writer, tend to lead to observation groups of very different sizes. Typically, there are a few very large groups, and many very small groups (often singletons), which again is challenging from a mathematical point of view.
- A considerable number of numerical variables in linguistics are related to each other in non-linear ways (as will also be illustrated in several studies in this book). Also, measurements can show autocorrelation patterns. This is a particular concern in psycholinguistics experiments in which measurements constitute time series, but it can also be an issue in corpus data where, for instance, what happened earlier in a conversation can affect what happens later. Therefore, techniques are needed that can deal with non-linear relations and with autocorrelated patterns.
- The ever-increasing size of linguistic corpora, some of which now have clearly entered the era of big data, as well as the vast amounts of data generated in modern psycholinguistics labs, can lead to huge data sets. Applying regression techniques to such huge data sets, especially when the models are complex, can introduce computational issues.

These are some of the issues addressed in this book.

3 Mixed Models in This Book

This book offers a broad window on the use of mixed models in linguistics, with different chapters zooming in on different subfields of linguistics. Chapter 2, while not specifically discussing linguistic examples, zooms in on the analysis of huge data sets, discussing solutions that can be of great value for the analysis of the type of huge data sets that are often encountered in modern corpus linguistics. Chapter 3 zooms in on applications in second language acquisition and language and education. In Chap. 4, we look at examples from phonology, psycholinguistics and neurolinguistics. In Chap. 5, dialectometric and sociolinguistic data are discussed. In Chaps. 6 and 7, finally, examples from corpus linguistics are discussed.

Throughout the chapters, we encounter very different types of random effect structures. In Chap. 2, we look at the typical type of hierarchical structures of 'individuals within groups' that is often encountered in studies on language and education (e.g. pupils within classes within schools). In Chap. 3, the random-effect factors are the mother tongue and the second language of individuals who study Dutch as a third language. In Chap. 4, random-effect factors are participants and words. In Chap. 5, random-effect factors are speakers, words, and locations. In Chaps. 6 and 7, finally, random-effect factors are words.

1 Introduction

The different chapters address different important issues related to the use of mixed models in linguistics. For instance, if we briefly revisit the issues that were listed in the previous section, we see that the issue of the sometimes difficult borderline between fixed-effect and random-effect factors plays a role both in Chap. 3 (mother tongue and second language) and Chap. 7 (genre). The specific difficulties of using and interpreting mixed-effect logistic regression analysis are specifically addressed in Chaps. 1 and 6. Non-linear relations are discussed in Chaps. 4 and 5, and autocorrelation patterns are addressed in Chap. 4. Huge data sets, finally, are specifically addressed in Chap. 2.

4 Software Used in the Book

The goal of this book is to not only discuss mixed models at a conceptual level, but to also discuss the practical usage of the technique. Nowadays, many different statistical packages, including all major commercial tools, offers good support for mixed models. In linguistics too, many different tools are being used for running mixed models. That being said, it is probably fair to say that at present, for many linguists the statistical software environment R is the tool of choice for conducting mixed-effects regression analyses. This tendency is also reflected in this book. Most chapters in this book provide R code for the types of analyses that are being discussed. The authors provide this code either by including the most important pieces of R code in the text, or by providing a URL where an R script or an R paper package can be downloaded.

5 Chapters in This Book

To conclude this first chapter, we will introduce the chapters in this book in a bit more detail.

In Chap. 2, Verbeke et al. present an introduction to mixed models (using the alternative term *clusters* to refer to grouped observations) that specifically focuses on the correct interpretation of the parameters in the models, and on possible pitfalls and misunderstandings. For instance, they illustrate that in a logistic mixed model, a type of model that is very often used in linguistics, fixed effects no longer have a population-average interpretation (as they do in linear mixed models). Instead of describing average trends in the population (across clusters), they describe trends in average clusters. Next, they illustrate that Wald tests, likelihood ratio tests, and score test statistics cannot straightforwardly be used to test whether between-cluster variability is significant (i.e. to test whether a certain random effect is needed in a model), but that instead corrections are needed (often using mixtures of χ^2 distributions). Also, they show that the distribution of empirical Bayes predictions for random effects (the so-called BLUPS) should not be used to test distributional assumptions made in the model. They also discuss pseudo-likelihood techniques, which enable the researcher to analyze data sets that are so large that standard likelihood based inference is no longer feasible, and which therefore, in light of the ever growing size of linguistic corpora and other linguistic data collections, offer a most welcome addition to the linguist's toolset. Most examples in Chap. 2 illustrate cases with nested random effects. In the next chapter, Chap. 3, focus shifts to crossed random effects.

In Chap. 3, Schepens et al. illustrate that crossed random effects may have more complex interrelationships than is often assumed. Focus is on model selection (specifically the selection of the most appropriate random effects structure), where the authors specifically recommend that researchers compare the fit of their crossed random effects models with the fit of models that also include the respective random interaction effects. The chapter reports on a study, based on a large state examination database, of the effect of L1 (mother tongue) and L2 (second language) on the proficiency in Dutch as an L3. L1 and L2 are treated as two crossed sources of random variation. The authors want to inspect whether and how the variation across the levels of one random-effect factor (e.g. L2) depends on the levels of another random-effect factor (e.g. L1). For example: could it be that L1 Spanish learners benefit more from L2 English than L1 German learners do? One way of investigating this type of interrelatedness between random effects, which the authors claim to often be an issue in observational studies, is to incorporate an x-by-y random interaction effect, where x and y are the crossed random effects. The sample used for this study has data for 73 L1s, 44 L2s (one of which is the value 'none'), and 759 L1–L2 combinations. Model selection, as far as the random effect structure is concerned, consisted of the comparison of four models. The first model is a model with a random intercept for L1–L2 (Model 1). The second model is a model with crossed random intercepts for L1 and L2 (Model 2). Next, a model with random intercepts for both L1 and L1–L2 is inspected (Model 3). Finally, a model with the crossed random intercepts for L1 and L2, as well as an additional random intercept for the interaction effect L1–L2 is inspected (Model 4). Additionally, all these models contain the same range of fixed effects, as well as an additional random intercept for 'country of birth'. The authors argue that likelihood ratio tests and inspection of the estimated parameters indicate that Model 4 explains the data significantly better than the other models, with a larger proportion of variance being attributed to L1 factor than to L2 factors. The authors offer a detailed illustration of carefully executed model selection.

In Chap. 4, Baayen et al. address the case of responses constituting time series, which is quite common in experimental data in the field of linguistics. This situation may raise the problem of autocorrelated errors, a problem which in turn can potentially lead to anti-conservatism of p-values as well as to a more blurred window on the quantitative structure of the data. The paper illustrates two tools offered by generalized additive mixed models (gamms), as implemented by the R package mgcv, for dealing with autocorrelated errors. Generalized additive mixed models extend the generalized linear mixed model with a large array of tools for modeling nonlinear dependencies between a response variable and one or more numeric predictors. The first tool in mgcv that can help us account for autocorrelated errors is the incorporation in the model of a first-order autoregressive process for the errors, which uses an autocorrelation parameter ρ. The second tool is the use of factor smooths for random-effect factors. These smooths are set up (by means of penalization) to yield the non-linear equivalents of random intercepts and random slopes in the classical linear framework.

Three cases studies are discussed. The first case study is on a word naming task; in this kind of task, participants are asked to respond to stimuli that are presented sequentially, so measurements for each participant result in a time series, and possibly a participant's later responses are not independent from his earlier responses. A model with a random intercept for verb (i.e. the word) and by-subject wiggly penalized curves for trial (i.e. position in the time series), in combination with a rather modest autoregressive parameter ρ of 0.3, is shown to almost completely account for the autocorrelation in the residuals. The second case study is on the pitch contour in the pronunciation of English three-constituent compounds. In this study, there are $12 \times 40 = 480$ elementary time series (viz. all combinations of 12 speaker and 40 English three-constituent compounds); in each elementary time series, pitch is measured at 100 moments in normalized time. The autocorrelation patterns that are much stronger than those in the first case study. Out of the three models the authors compare for this second case study, a model with by-compound and by-participant random wiggly curves as well as a high autoregressive parameter ρ of 0.98, is found to offer the best fit for the data and to remove most of the autocorrelation from the model residuals. The third case study models amplitude over time of the brain's electrophysiological response to visually presented compound words (EEG data). Focus is again on the complex random structure of the data and on the autocorrelation structure in the model residuals.

Throughout the three cases studies, the authors discuss model selection in detail. They illustrate the different ways in which the introduction of random curves and of an autoregressive parameter ρ can impact the models. Regarding the latter, they more specifically argue and illustrate that, when residuals reveal autocorrelational structure, ρ should be chosen high enough to remove substantial autocorrelational structure, but not so high that new, artificial autocorrelational structure is artefactually forced onto the data.

In Chap. 5, Wieling et al. investigate which factors influence the linguistic distances of Catalan dialectal pronunciations from standard Catalan. Using a large data set of catalan dialect pronunciation of 357 words by 320 speakers of varying age coming from 40 locations, the authors show that the speakers of Catalan in Catalonia and Andorra use a variety of Catalan that is closer to the standard than the variety spoken by speakers from Aragon. Because this tendency is particularly strong among younger speakers, they argue that the difference is at least in part due to the introduction of Catalan as an official language in the 1980s in Catalonia and Andorra but not in Aragon. As far as design is concerned, their study adopts a dialectometric approach that is enriched with social factors. More specifically, their study is dialectometric, in the sense that they aggregate over many linguistic variables, but unlike many dialectometric studies, they do incorporate age group (which gives them a window on linguistic change) and several other social factors. The response variable in their model is pronunciation distance from the standard pronunciation, operationalized as (log-transformed and centered) normalized PMI-based Levenshtein distance. They use a generalized additive mixed-effects regression model in which geography is modeled by a non-linear interaction of longitude and latitude, and in which additionally location is included as a random-effect factor to capture location-based effects not captured by geographic coordinates. Fixed-effect predictors include word-specific variables, as well as location-specific and speaker-related social variables. Model selection is discussed in detail. The authors argue in favor of using both random intercepts and random slopes, in order to avoid anti-conservative p-values. They also argue in favor standardization of predictors. They also touch upon the issue of whether or not to use a maximally complex random-effects structure (see [3, 4]).

In Chap. 6, Barth and Kapatsinski address an issue that is very common in corpus data. If we use *word* as a random factor in the specific context of corpus data, we are faced with two specific properties of natural language: (1) word frequency distributions are such that in observational data such as corpus data, a small number of words will have exceptionally high frequencies; (2) at least according to some linguistics theories (most notably lexical diffusion theory), high frequency words tend to behave differently from other words (e.g. more articulatory reduction and semantic bleaching, and more retention of grammatical patterns that are no longer productive). The authors use simulation and cross-validation tests to investigate what are the implications of this situation for the role of random factors such as *word* in quantitative corpus linguistics research, and how they should be dealt with for the purpose of gauging fixed effects, selecting models and establishing model quality. First of all, they argue that specifically in the case of corpus-like sampling (which is bound to be unbalanced with respect to word frequencies), the inclusion of random effects is needed to obtain accurate fixed-effect coefficients. They show that corpus-like (unbalanced) sampling greatly diminishes the predictive power of fixed-effects-only models; it also hurts mixed-effects models, but to a much lesser extent. Second, they argue that whereas random factors need to be included in the models in order to more accurately capture the fixed effects in the model, at the same time it is the predictiveness of the fixed effects only that should guide model selection. In other words, they argue against using the fit (in their case, the concordance index C) of the complete model (including random effects) to evaluate mixed models, as is often done in linguistic research. Instead, evaluation of the fit of the model should be done by examining how much variance is captured by the fixed effects alone. Extrapolating their findings to the measures introduced in [9], they advocate using marginal R^2 rather than conditional R^2. This chapter can also be read as an argument in favor of cross-validation of regression models, which is a practice that is known in linguistics, but definitely is not common.

In Chap. 7, Zhang et al. present a mixed-effects logistic regression analysis by means of which they model how, in newspaper data and online forum data in Mainland Chinese and Taiwan Chinese, people choose between using either a literal or a metonymic expression when they refer to a government. Literal expressions included in the study are two Chinese words that are the Chinese counterparts of the English words *government* and *authorities* respectively. Metonymic expressions included in the study are all usages of a PLACENAME FOR GOVERNMENT metonymy encountered in the data (typically country names, or names of capitals or official residences of state leaders or governments). The source of random variation in this study is the main verb of the sentence in which the reference to a government occurs (often, but not always, with this government being the subject of this verb). Fixed-effect predictors in the model include conceptual, grammatical/discursive and lectal variables. The study illustrates that the choice of literal vs. metonymic expressions is the result of a complex interplay of these three types of variables. Most notably, contexts favoring the use of a PLACENAME FOR GOVERNMENT metonymy are the discussion of 'general topics of global importance' (as opposed to domain-specific topics and/or topics of only local importance), the syntactic role of the government being the subject of the sentence, and the situation of the reference to the government featuring in the title of the text (rather than in the body). Two models are being presented and discussed. First, a 'global' model is fit for all data points (including references to many different governments). Then, a second model is fit for the subset of only those observations that refer to the MAINLAND CHINESE GOVERNMENT. An interesting issue regarding the random structure in the data of this study, is the question where to draw the line between what are random-effect and what are fixed-effect factor. Two factors in this paper, viz. lect/variety/style and topic, are factors that in linguistics often are treated as random-effect factors, but that here, in light of the specific research goals (specifically, the lack of desire to extrapolate beyond the topics and lects studied here) are treated as fixed-effect factors.

Acknowledgements The idea for this book arose at the 2012 Leuven Statistics Days conference, the theme of which was "Mixed models and modern multivariate methods in linguistics" (http://lstat.kuleuven.be/research/lsd/lsd2012/index.htm). The conference took place at the KU Leuven and was co-organized by LStat (Leuven Statistics Research Centre) and the linguistic research group QLVL. We thank all conference participants for their contributions to the conference. We also thank all authors for contributing to this book, and we thank all anonymous referees for their important criticisms.

References

1. Baayen RH (2008) Analyzing linguistic data: a practical introduction to Statistics using R. Cambridge University Press, Cambridge
2. Baayen RH, Davidson DJ, Bates DM (2008) Mixed-effects modeling with crossed random effects for subjects and items. J Mem Lang 59(4):390–412
3. Baayen RH, Vasishth S, Bates D, Kliegl R (2015) Out of the cage of shadows. arxiv.org. http://arxiv.org/abs/1511.03120
4. Barr DJ, Levy R, Scheepers C, Tily HJ (2013) Random effects structure for confirmatory hypothesis testing: keep it maximal. J Mem Lang 68(3):255–278

5. Bresnan J, Cueni A, Nikitina T, Baayen RH (2007) Predicting the dative alternation. In: Bouma G, Kraemer I, Zwarts J (eds) Cognitive foundations of interpretation. Royal Netherlands Academy of Arts and Sciences, Amsterdam, pp 69–94
6. Dijkstra T, Miwa K, Brummelhuis B, Sappelli M, Baayen RH (2010) How cross-language similarity and task demands affect cognate recognition. J Mem Lang 62:284–301
7. Jaeger F (2008) Categorical data analysis: away from ANOVAs (transformation or not) and towards logit mixed models. J Mem Lang 59(4):434–446
8. Johnson DE (2009) Getting off the GoldVarb standard: introducing Rbrul for mixed-effects variable rule analysis. Lang Linguist Compass 3:359–383
9. Nakagawa S, Schielzeth H (2013) A general and simple method for obtaining R^2 from generalized linear mixed-effects models. Methods Ecol Evol 4:133–142
10. Wieling M, Nerbonne J, Baayen RH (2011) Quantitative social dialectology: explaining linguistic variation geographically and socially. PLoS ONE 6(9):e23613

Chapter 2
Mixed Models with Emphasis on Large Data Sets

Geert Verbeke, Geert Molenberghs, Steffen Fieuws, and Samuel Iddi

Abstract In many contexts, hierarchical, multilevel, or clustered data are collected. Examples are longitudinal studies in which subjects are measured repeatedly at various time points (measurements within subject), surveys in which all members of a sample of families are questioned (members within families), educational data in which students from various schools are tested (students within schools), etc. From a statistical perspective, the challenge is to account for the fact that the measurements within clusters are not necessarily independent anymore, implying that standard models such as linear regression or generalized linear regression are no longer applicable. Mixed models are currently amongst the most flexible models for the analysis of such data. They can be interpreted as standard linear, generalized linear, or non-linear models, with cluster-specific random effects shared by all measurements within the cluster, hereby implicitly accounting for within-cluster associations. In this chapter, mixed models will be introduced with special attention for the correct interpretation of the parameters in the models. Also, examples will be given of situations in which results obtained from fitting mixed models are incorrectly interpreted. Many commercial software packages nowadays include mixed model procedures. However, when (extremely) large data sets are to be analyzed, standard likelihood based inference is no longer feasible. Examples include data sets with crossed random effects, with many clusters, with many observations per cluster, or contexts where mixed models are used to build a joint model for high-dimensional multivariate responses. In such cases, pseudo-likelihood techniques provide good alternatives. Various versions will be presented and illustrated. All concepts will be introduced and extensively illustrated using data sets from various contexts.

G. Verbeke (✉) · G. Molenberghs · S. Fieuws · S. Iddi
I-BioStat, Katholieke Universiteit Leuven, B-3000 Leuven, Belgium

I-BioStat, Universiteit Hasselt, B-3590 Hasselt, Belgium
e-mail: geert.verbeke@med.kuleuven.be; geert.molenberghs@uhasselt.be;
steffen.fieuws@med.kuleuven.be; samuel.iddi@med.kuleuven.be

© Springer International Publishing AG 2018
D. Speelman et al. (eds.), *Mixed-Effects Regression Models in Linguistics*, Quantitative Methods in the Humanities and Social Sciences, https://doi.org/10.1007/978-3-319-69830-4_2

1 Introduction

In many contexts, hierarchical, multilevel, or clustered data are collected. Examples are longitudinal studies in which subjects are measured repeatedly at various time points (measurements within subject), surveys in which all members of a sample of families are questioned (members within families), educational data in which students from various schools are tested (students within schools), etc. From a statistical perspective, the challenge is to account for the fact that the measurements within clusters are not necessarily independent anymore, implying that standard models such as linear regression or generalized linear regression are no longer applicable.

Mixed models are currently amongst the most flexible models for the analysis of such data. They can be interpreted as standard linear, generalized linear, or non-linear models, with cluster-specific random effects shared by all measurements within the cluster, hereby implicitly accounting for within-cluster associations. In this chapter, mixed models will be introduced with special attention on the correct interpretation of the parameters in the models. Also, examples will be given of situations in which results obtained from fitting mixed models are incorrectly interpreted.

Many commercial software packages nowadays include mixed model procedures. However, when (extremely) large data sets are to be analyzed, standard likelihood based inference is no longer feasible. Examples include data sets with crossed random effects, with many clusters, with many observations per cluster, or contexts where mixed models are used to build a joint model for high-dimensional multivariate responses. In such cases, pseudo-likelihood techniques provide good alternatives. Various versions will be presented and illustrated.

All concepts will be introduced and extensively illustrated using data sets from various contexts. The structure of the chapter is as follows. In Sect. 2, an introduction is given to linear as well as generalized linear mixed models. Both model families will first be introduced in the context of a particular data set, followed with some brief discussion of estimation and inference. An extensive case study will be presented in Sect. 3, where one of the examples from Sect. 2 will be analysed in full detail, illustrating issues practicing statisticians are often confronted with when applying mixed models. When mixed models are to be used for the analysis of large data sets, computational issues may arise. Section 4 provides a split-sample solution based on pseudo-likelihood inference, and several examples are discussed. Finally, an overall conclusion is presented in Sect. 5.

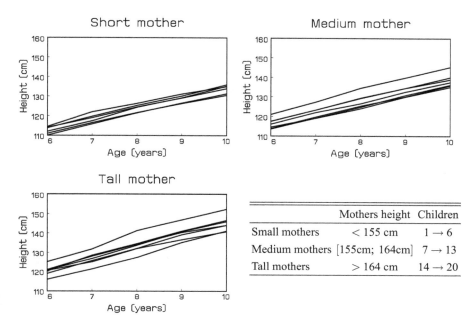

Fig. 2.1 Heights of schoolgirls. Growth curves of 20 school girls from age 6 to 10, for girls with small, medium, or tall mothers

2 Mixed Models

2.1 Linear Mixed Models

Goldstein [12] reports growth curves of 20 preadolescent girls, measured on a yearly basis from age 6 to 10. The girls were classified according to the height of their mother. The individual profiles are shown in Fig. 2.1, for each group separately. The measurements are given at exact years of age, some having been previously adjusted to these. The values Goldstein reports for the fifth girl in the first group are 114.5, 112, 126.4, 131.2, and 135.0. This suggests that the second measurement is incorrect. We therefore replaced it by 122. Of primary interest is to test whether the growth of these schoolgirls is related to the height of their mothers. An extensive analysis of this data set can be found in Section 4.2 of [29]. Here, we will use the data to introduce the linear mixed model on an intuitive basis.

Graphical exploration of Fig. 2.1 suggests that the evolution of each child can be well described by a linear function over time, but with subject-specific intercepts and possibly subject-specific slopes as well. Let Y_{ij} denote the jth measurement for the ith child, taken at time t_j (age), $i = 1, \ldots, 20, j = 1, \ldots, 5$. It is then assumed that Y_{ij} can be modeled as

$$Y_{ij} = \beta_{1i} + \beta_{2i} t_j + \varepsilon_{ij}, \tag{2.1}$$

in which β_{1i} and β_{2i} are the intercept and slope of child i, respectively, and where the terms ε_{ij} represent the traditional errors, assumed to be normally distributed with mean zero and variance σ_{res}^2. This residual variance expresses the fact that the observations do not perfectly meet the linearity assumption. From this perspective, σ_{res}^2 can be interpreted as representing unexplained within-subject variability. Obviously, the intercepts β_{1i} are different for all subjects, and this variability can be partially explained by the fact that the children belong to different groups. Let S_i, M_i, and T_i represent indicators for group membership (short–medium–tall), then a possible model to explain variability between the children in terms of their intercepts is

$$\beta_{1i} = \beta_1 S_i + \beta_3 M_i + \beta_5 T_i + b_{1i}. \tag{2.2}$$

The parameters β_1, β_3, and β_5 represent the average intercepts in the short mother group, the medium mother group, and the tall mother group. Obviously, a similar model can be used for the child-specific slopes β_{2i}, leading to

$$\beta_{2i} = \beta_2 S_i + \beta_4 M_i + \beta_6 T_i + b_{2i}, \tag{2.3}$$

where β_2, β_4, and β_6 now represent the average slopes in the three groups, respectively. Note that the error components b_{1i} and b_{2i} indicate that children within the same group do not necessarily have the same intercepts or slopes. Hence, the b_{1i} and b_{2i} represent unexplained between-subject variability, and they are assumed to be jointly normally distributed. More specifically, it is assumed that $\boldsymbol{b}_i = (b_{1i}, b_{2i})' \sim N(0, D)$ for some 2×2 covariance matrix D, with entries d_{pq}, $p, q = 1, 2$.

Combination of Eq. (2.1) with Eqs. (2.2) and (2.3) yields

$$Y_{ij} = (\beta_1 S_i + \beta_3 M_i + \beta_5 T_i + b_{1i}) + (\beta_2 S_i + \beta_4 M_i + \beta_6 T_i + b_{2i}) t_j + \varepsilon_{ij}, \tag{2.4}$$

which is a linear regression model in which some parameters are the same for all subjects (*fixed effects*), while other parameters are subject-specific and assumed random (*random effects*), resulting in a so-called mixed-effects model, often briefly termed *mixed model*.

As mentioned in Sect. 1, mixed models provide a natural way to allow for within-cluster associations. For example, model (2.4) implies that two observations Y_{ij} and Y_{ik} of child i have covariance

$$\text{cov}(Y_{ij}, Y_{ik}) = d_{22} t_j t_k + d_{12}(t_j + t_k) + d_{11} + \sigma_{res}^2,$$

clearly showing that observations belonging to the same child are no longer assumed independent, as classical linear regression models would have assumed. Note also that, because all random terms in (2.4) have mean zero, the expectation of Y_{ij} equals

$$E(Y_{ij}) = (\beta_1 S_i + \beta_3 M_i + \beta_5 T_i) + (\beta_2 S_i + \beta_4 M_i + \beta_6 T_i) t_j,$$

showing that the fixed effects are modelling *systematic trends* in the data, while random effects are used to model the *association structure*.

2.2 Generalized Linear Mixed Models

Expression (2.4) is an example of a *linear* mixed model, which is any linear regression containing fixed as well as random regression parameters. When the outcome of interest is not continuous, a model such as (2.1) is not necessarily meaningful anymore, and needs to be replaced by a more appropriate model. As an example, we consider data from the Leuven Diabetes Project [2, 3, 11], in which the impact is studied of offering general practioners (GP's) assistance of a diabetes care team, consisting of a nurse educator, a dietician, an ophthalmologist, and an internal medicine doctor, for the treatment of their diabetes patients. GP's from the area of Leuven (Belgium) were randomly assigned to either a low intervention program or a high intervention program. Here, we consider the data from the high intervention program only, resulting in a total of 1577 patients, treated by 61 GP's. The number of patients per GP ranged from 5 to 138, with median value 47. Patients were scheduled to be measured twice, once when the program was initiated, and once after one year in the study. One of the outcomes of primary interest was HbA1c, glycosylated hemoglobin, a molecule in red blood cells that attaches to glucose (blood sugar), high values reflecting more glucose in blood. HbA1c indicates how well diabetes has been managed over the last 2 or 3 months. Non-diabetics have values between 4 and 6%. HbA1c above 7% indicates that diabetes is poorly controlled, implying higher risk for long-term complications. Here we will consider a dichotomized version of HbA1c, defined as $Y = 1$ if HbA1c is less than 7% and 0 otherwise.

The binary nature of the outcome no longer allows the use of a linear model as in (2.4). Instead, a logistic model will be used, in which the probability $P(Y = 1)$ is modeled on a logit scale. Note also that, in contrast to the growth curves in Sect. 2.1, an additional hierarchy is present in the data. Indeed, patients are clustered within GP's, and repeated (longitudinal) measurements are clustered within patients. This will be accounted for by random effects at two different levels. Random effects at the GP level will account for the clustering of patients within GP's, while random effects at the patient level will account for the clustering of the repeated measures within patients. More specifically, let Y_{ijk} denote the outcome at time point $t_k = 0, 1$, for patient j treated by GP i, $i = 1, \ldots, 61$, $j = 1, \ldots, n_i$, it is assumed that

$$P\left[Y_{ijk} = 1 | a_i, b_{j(i)}\right] = \frac{\exp\left[\beta_0 + \beta_1 t_k + a_i + b_{j(i)}\right]}{1 + \exp\left[\beta_0 + \beta_1 t_k + a_i + b_{j(i)}\right]}. \quad (2.5)$$

As before, the parameters β_0 and β_1 model the systematic trend in the population, while the parameters a_i and $b_{j(i)}$ model the variability between the GP's and between the patients, respectively. The notation $j(i)$ is used to explicitly denote that patients are nested within GP's. As before, cluster (GP or patient) specific parameters a_i and $b_{j(i)}$ are assumed normally distributed with means 0 and variances σ^2_{GP} and σ^2_{PAT}, respectively. Note that our notation now explicitly acknowledges the fact that the model is conditional on a particular GP and a particular patient treated by that GP (probability conditional on a_i and $b_{j(i)}$).

Model (2.5) is an example of a logistic mixed model, which is one particular example of a *generalized linear mixed model*, useful for the analysis of binary outcomes. Other examples include proportional odds mixed models for ordinal outcomes or Poisson mixed models for count data. Examples can be found in [15].

2.3 Estimation and Inference

Most commercially available software packages nowadays allow fitting of linear and generalized linear mixed models. In general, unless a fully Bayesian approach is followed (see, e.g., [10]), estimation and inference are based on likelihood principles. Assuming clusters to be independent, the likelihood can easily be constructed and maximized. Note however that the models often contain many parameters (fixed effects and random effects). Direct maximization therefore may require maximization over very high dimensional parameter spaces. Furthermore, as the number of parameters increases with the number of clusters, classical asymptotics do not hold anymore. To avoid these problems, likelihood estimation is based on the marginal likelihood, i.e., the likelihood obtained from integrating over the random effects which are assumed normally distributed. The resulting marginal models depend on the fixed effects and the covariance parameters only, the number of which is relatively small compared to the original number of parameters in the models.

Except in simple cases such as the linear mixed model, integrating out the random effects is not possible analytically, and approximations are needed. Many solutions have been proposed in the statistical literature, including Taylor series expansions, approximations of the data, and numerical approximations of the integrals. Details on the many available estimation methods, together with illustrations and comparisons, can be found in [15, 18, 30].

Because fitting of (generalized) linear mixed models is based on maximum likelihood principles, inferences for the parameters are readily obtained from classical maximum likelihood theory. Indeed, assuming the fitted model is appropriate, the obtained estimators are asymptotically normally distributed with the correct values as means, and with the inverse Fisher information matrix as covariance matrix. Hence, asymptotic Wald-type tests, comparing standardized estimates to the standard normal distribution can be easily performed. Alternatively, asymptotic likelihood ratio tests and score tests can be used as well. One complication often ignored in practice is that estimation of parameters in covariance structures such as the random-effects covariance D occurs under the restriction of non-negative definiteness for the resulting covariance structure. This implies that null hypotheses involving such parameters are often on the boundary of the parameter space. As a result, classical maximum likelihood theory no longer applies, and none of the classical testing procedures (Wald, likelihood ratio, score) remain valid. This will be further discussed in Sect. 3, in the context of the Leuven Diabetes Project.

While interest is often primarily in estimation and inference for the fixed effects and/or the variance components only, one sometimes is interested in obtaining predictions for the random effects as well. They reflect between-cluster variability, which makes them helpful for detecting special profiles (i.e., outlying individuals) or groups of individuals showing extraordinary behavior. For example, in the context of the Diabetes Project Leuven, interest might be in predicting the GP effects a_i to assess the performance of each individual GP. Usually random effects in mixed models are predicted using their *empirical Bayes (EB)* estimate. The EB estimate of a random effect is its posterior mode, i.e., the most likely value for the random effect, conditional on the observed data for that particular cluster. In the Diabetes Project Leuven, the EB estimate of the contribution $b_{j(i)}$ for patient j treated by GP i would be the most likely value for $b_{j(i)}$ given the data that have been observed for that particular patient. The EB estimate of the contribution a_i for GP i would be the most likely value for a_i given the observed data of all patients treated by that particular GP.

3 Mixed Models in Action: The Leuven Diabetes Project

While (generalized) linear mixed models can be interpreted as relatively straightforward extensions of the traditional (generalized) linear models [13] to the context of clustered data, the interpretation of some of the results is less straightforward and needs special attention. Scientists applying mixed models are not always aware of these peculiarities, leading to incorrect interpretations of some of the results obtained. A number of such situations will be described in this section, in the context of the Leuven Diabetes Project introduced in Sect. 2.2, and analysed using model (2.5).

3.1 Interpretation of the Fixed Effects

We first consider inference for the fixed effects β_0 and β_1. Estimates, associated standard errors and p-values are reported in Table 2.1. As explained in Sect. 2.1, fixed effects in linear mixed models can be interpreted as average intercepts and slopes, hence modeling the average trends in the population. Note however, that

Table 2.1 Estimates, associated standard errors, Z-statistics, and p-values for the fixed effects in model (2.5)

Effect	Estimate (se)	Z	p-value
Intercept β_0	0.1662 (0.0796)	2.0879	0.0368
Time β_1	0.6240 (0.0812)	7.6847	<0.0001

this interpretation is based entirely on the linearity of model (2.1). In other models, this interpretation no longer holds. For example, the average trend implied by the logistic mixed model (2.5) used for the analysis of the Leuven Diabetes Project is obtained from calculating

$$E\{P[Y_{ijk} = 1 | a_i, b_{j(i)}]\} = E\left\{\frac{\exp[\beta_0 + \beta_1 t_k + a_i + b_{j(i)}]}{1 + \exp[\beta_0 + \beta_1 t_k + a_i + b_{j(i)}]}\right\}, \quad (2.6)$$

where the expectation is over the random effects a_i and $b_{j(i)}$, both assumed to be normally distributed. Although the random effects have mean zero, the non-linear relation between $P[Y_{ijk} = 1 | a_i, b_{j(i)}]$ and the random effects, implies that the expectation in (2.6) cannot be obtained by simply replacing the random effects by zero, i.e., the average trend in the population is not given by

$$P[Y_{ijk} = 1 | a_i = 0, b_{j(i)} = 0] = \frac{\exp[\beta_0 + \beta_1 t_k]}{1 + \exp[\beta_0 + \beta_1 t_k]} \quad (2.7)$$

Hence, in contrast to the linear mixed model, the fixed effects no longer have a population-average interpretation. Instead, they describe the evolution of the *average patient* treated by the *average GP*, i.e., a patient and GP with random effects equal to zero, which surprisingly is not the average evolution in the population.

This phenomenon can most easily be understood in a graphical way. Figure 2.2 shows the evolution of the probability $P(Y = 1)$ for 20 randomly selected patients (thin lines), based on model (2.5). All patients show a logistic evolution with the same slope β_1 and with intercepts $\beta_0 + \beta_1 t_k + a_i + b_{j(i)}$, which are subject-specific, depending on the actual patient under consideration, and the actual GP who is treating that patient. The population-average trend however, is the same probability but now averaged over the entire population. It is obtained from averaging the subject-specific probabilities at each time point, and this results in the bold line in Fig. 2.2. Clearly, the average trend is less steep than each of the individual curves. Furthermore, the larger the between-GP and between-patient variances σ_{GP}^2 and σ_{PAT}^2 are, the more different the individual trends in Fig. 2.2, leading to a more severe

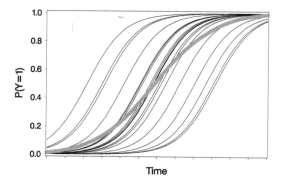

Fig. 2.2 Graphical representation of the logistic mixed model (thin lines), with indication of the population-average trend (bold line)

deviation of the slope of the population-average trend from the individual trends. This clearly illustrates that the fixed effects in generalized linear models should not be interpreted as effects describing the average trends in the population. Instead, they describe the trend of the average cluster, i.e., a cluster with random effect(s) equal to zero.

3.2 Tests for Variance Components

Although fixed effects in generalized linear mixed models, in general, should not be interpreted as the average effect of covariates of interest (such as time), they do describe systematic trends and therefore often are of primary interest. This notwithstanding, researchers sometimes are interested in inferences for the variance components as well. In the Leuven Diabetes Project, for example, one might be interested in assessing the amount of variability between GP's. Indeed, a large between-GP variance σ_{GP}^2 suggests that there exist substantial differences between GP's in terms of being able to control diabetes of their patients. In such cases, additional patient- or GP-specific covariates may be added to the model in an attempt to explain the large between-GP variability. Table 2.2 shows estimates, associated standard errors, Z-statistics, and p-values for the two variance components in model (2.5). Clearly, the between-patient variability is much larger than the between-GP variability.

Testing whether the between-GP variability is significant is equivalent to testing the null hypothesis $H_0 : \sigma_{GP}^2 = 0$. Naively, one might argue that the asymptotic normality of the maximum likelihood estimates allows testing the hypothesis using a standard Wald test, based on a Z-statistic equal to the estimate divided by the associated standard error. In our example, the Z-statistic equals 2.6496 leading to a p-value equal to $p = 0.0081$. However, as the null hypothesis tested is not in the interior of the parameter space, the traditional asymptotic properties of maximum likelihood estimates no longer hold, implying that Wald test, likelihood ratio tests, and score test statistics no longer follow the asymptotic distribution traditionally used for the calculation of p-values, and corrections are needed. In the current example, the corrected p-values for testing $H_0 : \sigma_{GP}^2 = 0$ and $H_0 : \sigma_{PAT}^2 = 0$ are only half the original ones, but other corrections may be needed in other contexts.

As an example, consider a model as in Sect. 2.1 with random intercepts and slopes, and consider testing whether random slopes are needed in the model. Under

Table 2.2 Estimates, associated standard errors, Z-statistics, and p-values (incorrect → corrected) for the variances of the random effects in model (2.5)

Effect	Estimate (se)	Z	p-value
Between GP variance σ_{GP}^2	0.1399 (0.0528)	2.6496	0.0081 → 0.0041
Between patient variance σ_{PAT}^2	1.1154 (0.1308)	8.5275	<0.0001 → <0.0001

the alternative hypothesis, D is a 2×2 non-negative definite matrix, which reduces to a non-negative scalar under the null hypothesis. Clearly, the null hypothesis is on the boundary of the alternative parameter space given that it requires the random-slopes variance to be zero. Stram and Lee [25, 26] have shown that, in this case, the asymptotic null distribution for the likelihood ratio test statistic is a mixture of a χ_1^2 and a χ_2^2, with equal probability $1/2$, rather than the standard χ_2^2 one would expect under the classical likelihood theory. In general, the asymptotic null distribution for the likelihood ratio test statistic for testing a null hypothesis which allows for q correlated random effects versus an alternative of $q+1$ correlated random effects, is a mixture of a χ_q^2 and a χ_{q+1}^2, with equal probability $1/2$. For more general settings, e.g., comparing models with q and $q + q'$ ($q' > 1$) correlated random effects, the null distribution is a mixture of χ^2 random variables [21, 22], the weights of which can only be calculated analytically in a number of special cases. Similar results can be derived for the score test. The correction needed for testing $H_0 : \sigma_{GP}^2 = 0$ and $H_0 : \sigma_{PAT}^2 = 0$ in the Leuven Diabetes example followed from the results of [22]. Building upon Silvapulle and Silvapulle [24], Molenberghs and Verbeke [16], and Verbeke and Molenberghs [31, 32] have shown that the Wald test and the score test are asymptotically equivalent to the likelihood ratio test, and that the same mixtures of χ^2 distributions appear as asymptotic null distributions.

3.3 Empirical Bayes Estimation

As explained in Sect. 2.3, predictions of random effects are needed in cases where predictions for particular clusters are of interest, or if outlying clusters are to be identified. Figure 2.3 shows histograms of the EB estimates for the random GP effects a_i and random patient effects $b_{j(i)}$ in model (2.5) used in the analysis of the Leuven Diabetes Project. The histogram of predicted patient effects suggests three clusters of patients, with approximate cut-offs for $\hat{b}_{j(i)}$ equal to -0.6 and 0.1. Since

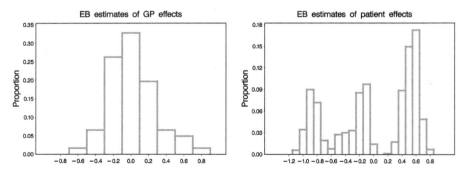

Fig. 2.3 Histograms of empirical Bayes estimates for the random GP effects a_i and random patient effects $b_{j(i)}$ in model (2.5)

Table 2.3 Cross-classification of observed response profiles with predicted patient effects, obtained from fitting model (2.5) to the Leuven Diabetes Project data

Y profile	$\hat{b}_{j(i)} < -0.6$	$-0.6 \leq \hat{b}_{j(i)} < -0.1$	$-0.1 \leq \hat{b}_{j(i)}$
$0 \longrightarrow 0$	345	0	0
$0 \longrightarrow 1$	0	275	0
$1 \longrightarrow 1$	0	0	677

the prediction $\hat{b}_{j(i)}$ is the most likely value for $b_{j(i)}$ given the observations available for that particular patient, and since each patient has at most two observations, it seems that the presence of the clusters is related to the very discrete nature of possible outcome profiles observed for the patients in this study. Indeed, patients with $Y_{ij1} = Y_{ij2} = 0$, i.e., $0 \to 0$, are expected to have (very) small predicted probabilities for reaching the target HbA1c less than 7%. Therefore, their prediction $\hat{b}_{j(i)}$ should be very small (negative). Patients with both outcomes equal to one, i.e., with profile $1 \to 1$, should have large predicted probabilities for reaching the target implying a prediction $\hat{b}_{j(i)}$ which is very large (positive). Patients with two different outcome values at both occasions, i.e., with profile $0 \to 1$ or $1 \to 0$, are expected to have intermediate predicted probabilities for reaching the target. Their prediction $\hat{b}_{j(i)}$ therefore should be of a moderate level. This can easily be confirmed from cross-tabulating the observed response profiles with the EB estimates for the patient effects, see Table 2.3. Note that all patients with HbA1c at target at the start of the study have their HbA1c at target one year later as well, i.e., no profiles $1 \to 0$ have been observed in this data set. The total number of patients in the table is less than the total number of patients present in the data set due to missing observations.

The very limited number of different patient profiles, which is a result of the binary outcome and the fact that at most two observations per patient are available is also clearly reflected in the scatterplot of predicted patient effects versus predicted GP effects, shown in Fig. 2.4. For each predicted GP effect, at most seven different predictions are obtained for patients treated by that particular GP. They correspond to the seven observed profiles $0 \to 0$, $0 \to 1$, $1 \to 1$, $0 \to \cdot$, $1 \to \cdot$, $\cdot \to 0$, and $\cdot \to 1$, in which $\cdot$ indicates the outcome value is missing. The negative linear trends in the scatterplot are also a side effect of the discrete nature of the outcomes. Indeed, consider two patients, j_1 and j_2, treated by different GP's, i_1 and i_2, but with the same response profile, e.g., $1 \to 1$. Their subject-specific models are given by

$$P\left[Y_{ijk} = 1 | a_{i_1}, b_{j_1(i_1)})\right] = \frac{\exp\left[\beta_0 + \beta_1 t_k + a_{i_1} + b_{j_1(i_1)}\right]}{1 + \exp\left[\beta_0 + \beta_1 t_k + a_{i_1} + b_{j_1(i_1)}\right]},$$

$$P\left[Y_{ijk} = 1 | a_{i_2}, b_{j_2(i_2)})\right] = \frac{\exp\left[\beta_0 + \beta_1 t_k + a_{i_2} + b_{j_2(i_2)}\right]}{1 + \exp\left[\beta_0 + \beta_1 t_k + a_{i_2} + b_{j_2(i_2)}\right]}.$$

Since both patients have the same data, their predicted probabilities should be the same at all time points, implying $\hat{a}_{i_1} + \hat{b}_{j_1(i_1)} = \hat{a}_{i_2} + \hat{b}_{j_2(i_2)}$. Hence, we expect the

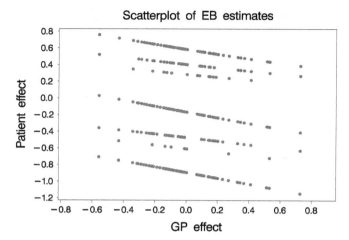

Fig. 2.4 Scatterplot of empirical Bayes estimates of random patient effects $b_{j(i)}$ versus random GP effects a_i in model (2.5)

sum $\hat{a}_i + \hat{b}_{j(i)}$ of GP and patient effects to be constant, implying a negative perfectly linear trend in the scatterplot of predicted patient effects and GP effects, for patients with the same outcome profile.

This example shows that, while the random effects in mixed models are assumed to be normally distributed, one should not necessarily expect EB predictions for those random effects to be normally distributed. The opposite holds as well. As an illustration, we report results from a small scale simulation of [28]. They simulated completely balanced continuous data from a linear random-intercepts model, with five longitudinal observations for 1000 subjects. The true random intercepts were sampled from a symmetric bimodal mixture of two normals, shown in the left panel of Fig. 2.5. Such a situation would arise when a factor with two levels is very predictive for the intercept but has not been included in the model. The data were analysed with a standard linear random-intercepts model assuming the random effects to be normally distributed. The right panel in Fig. 2.5 shows the EB predictions for the random effects. Obviously, the normality assumption forces the predictions to approximately satisfy normality, showing that EB predictions should not be used to test distributional assumptions made by the model. This suggests that, if interest is in studying the random-effects distribution, mixed models should be used with random-effects distributional assumptions that are sufficiently flexible [28, 30].

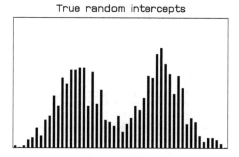
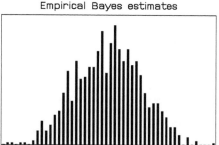

Fig. 2.5 Histograms of true (left panel) and predicted (right panel) random intercepts for simulated completely balanced continuous longitudinal data, with five longitudinal observations for 1000 subjects

4 Issues with Large Data Sets

Mixed models can nowadays easily be fitted using commercially available software packages. However, when large data sets need to be analysed, computational issues may occur. To focus ideas, consider data sets with N clusters and n measurements per cluster. Situations where large data sets arise include large observational studies (N large, e.g., [23]), examples from statistical genetics or functional data analysis (n large, e.g., [20, 27]), or large multivariate longitudinal studies (N as well as n large [5]). A graphical representation of such data is given in panel (a) of Fig. 2.6.

As discussed in Sect. 2.3, the fitting of mixed models implies integrating random effects for which approximations are needed. Because this needs to be done for each cluster in the data set separately, and at each step of the optimization process, this can become extremely time consuming in cases where the number N of clusters is (very) large. Alternatively, large numbers n of observations per cluster lead to high-dimensional multivariate distributions for the measurements per cluster. This creates numerical problems, even in the linear model, due to numerical inversion of large covariance structures. Molenberghs et al. [17] therefore proposed to split the entire sample in such a way that the calculations become feasible for each sub-sample. The model can then be fitted to the data of each sub-sample separately, and the results can be combined afterwards in an appropriate way. Inference follows from pseudo-likelihood ideas [1]. In Sect. 4.1, the general split-sample idea will be explained, and pseudo-likelihood inference will be briefly reviewed. Afterwards a number of ways to split large data sets will be discussed in Sect. 4.2.

4.1 The Split-Sample Idea

Suppose the original purpose was to use maximum likelihood estimation, and let $\ell(\boldsymbol{\theta})$ denote the log-likelihood function to be maximized with respect to the parameter vector $\boldsymbol{\theta}$, i.e., $\ell(\boldsymbol{\theta})$ equals

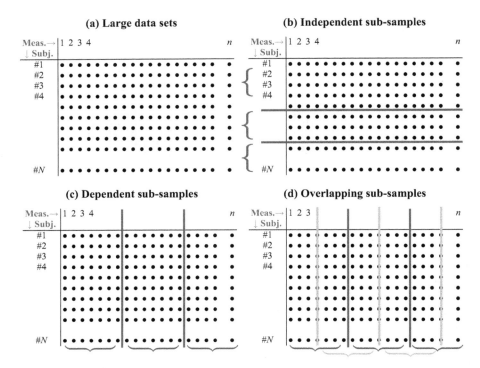

Fig. 2.6 Graphical representation of different ways to split large samples: (**a**) representation of a large sample, (**b**) split in independent samples, (**c**) split in dependent disjunctive samples, (**d**) split in overlapping samples. Braces indicate ways to split the sample in sub-samples

$$\ell(\boldsymbol{\theta}) = \sum_i \ell(\boldsymbol{y}_i|\boldsymbol{\theta}), \qquad (2.8)$$

where the summation is over the independent clusters in the data set, and where $\boldsymbol{y}_i$ is the vector of all observations available for cluster i. Pseudo-likelihood methodology replaces the log-likelihood contribution $\ell(\boldsymbol{y}_i|\boldsymbol{\theta})$ of cluster i in (2.8) by a weighted sum of log-likelihood contributions for sub-vectors $\boldsymbol{Y}_i^{(s)}$ of $\boldsymbol{Y}_i$. More specifically, rather than optimizing (2.8), the pseudo-log-likelihood function

$$p\ell(\boldsymbol{\psi}) = \sum_i \sum_s \delta_s \, \ell(\boldsymbol{y}_i^{(s)}|\boldsymbol{\psi}) \qquad (2.9)$$

is maximized with respect to $\boldsymbol{\psi}$, not necessarily identical to $\boldsymbol{\theta}$, for a well-chosen set of real numbers δ_s, chosen such that maximization of (2.9) becomes relatively straightforward, and such that $\boldsymbol{\psi}$ contains all parameters of interest. Although the so-obtained estimate $\hat{\boldsymbol{\psi}}$ is not the maximum likelihood estimate that would have been obtained if (2.8) had been maximized, it still has similar properties such as consistency and asymptotic normality. Asymptotic standard errors for the elements in $\hat{\boldsymbol{\psi}}$ are based on first- and second-order derivatives of $p\ell(\boldsymbol{\psi})$ with respect to $\boldsymbol{\psi}$. We refer to [1] for a full account of pseudo-likelihood inference.

4.2 Examples of How Large Data Sets Can Be Split

Depending on the context, the model under consideration, and the research questions, various ways can be explored to split the data set in sub-samples, all implying particular choices for the weights δ_s in (2.9). In cases where the number N of clusters is too large to analyse the entire data set at once, the most obvious split is to partition the clusters in M independent sets S_m of clusters, $m = 1, \ldots, M$. This situation is shown in panel (b) of Fig. 2.6. In each sub-sample the model is fitted, yielding an estimate $\hat{\boldsymbol{\theta}}_m$ of $\boldsymbol{\theta}$. This is equivalent to maximizing

$$p\ell(\boldsymbol{\psi}) = \sum_m \sum_{i \in S_m} \ell(\boldsymbol{Y}_i | \boldsymbol{\theta}_m), \qquad (2.10)$$

with respect to $\boldsymbol{\psi} = \{\boldsymbol{\theta}_1, \boldsymbol{\theta}_2, \ldots, \boldsymbol{\theta}_M\}$. Since (2.10) is a special case of (2.9), inference for $\boldsymbol{\psi}$ immediately follows from pseudo-likelihood theory. Note that, in this case, all $\boldsymbol{\theta}_m$ are equal to $\boldsymbol{\theta}$. Hence $\hat{\boldsymbol{\psi}}$ contains M independent estimators of $\boldsymbol{\theta}$. A single estimator is easily obtained from averaging all $\hat{\boldsymbol{\theta}}_m$ and asymptotic properties immediately follow.

In case of large n, the data may be partitioned in M sets S_m of clusters, $m = 1, \ldots, M$, as indicated in panel (c) of Fig. 2.6. Note that the sub-samples are no longer independent as each sub-sample S_m contains data of all clusters, and measurements from the same cluster are not necessarily independent. Let $\boldsymbol{Y}_i^{(m)}$ denote the observations in $\boldsymbol{Y}_i$ belonging to subsample S_m. Fitting the model to each subsample is now equivalent to maximizing

$$p\ell(\boldsymbol{\psi}) = \sum_m \sum_i \ell(\boldsymbol{Y}_i^{(m)} | \boldsymbol{\theta}_m), \qquad (2.11)$$

with respect to $\boldsymbol{\psi} = \{\boldsymbol{\theta}_1, \boldsymbol{\theta}_2, \ldots, \boldsymbol{\theta}_M\}$. Since (2.11) is a special case of (2.9), inference for $\boldsymbol{\psi}$ immediately follows from pseudo-likelihood theory. Note that, in this case, all $\boldsymbol{\theta}_m$ are not necessarily equal to $\boldsymbol{\theta}$. In a longitudinal context, for example, some parameters in $\boldsymbol{\theta}$ may characterize early evolutions therefore only appearing in the models for early observations in $\boldsymbol{Y}_i$, i.e., in some sub-vectors $\boldsymbol{Y}_i^{(m)}$ only. Appropriate combination of all $\hat{\boldsymbol{\theta}}_m$ into a single estimator for $\boldsymbol{\theta}$ very much depends on the precise model and data structure.

While the previous examples of the split-sample technique were based on partitioning the original sample in disjunctive sub-samples, some applications require overlapping sub-samples. Examples can be found in Fieuws and Verbeke [5, 6], Fieuws et al. [7, 9], and Fieuws, Verbeke, and Molenberghs [8], who analysed high-dimensional multivariate longitudinal data using mixed models. For each longitudinal outcome $Y_i^{(q)}$, $q = 1, \ldots, Q$ a mixed model was assumed with random effects $b_i^{(q)}$. Association between the longitudinal profiles for a single subject was accounted for by allowing the random effects to be correlated, leading to a large multivariate mixed model for $Y_i = \{Y_i^{(1)}, Y_i^{(2)}, \ldots, Y_i^{(Q)}\}$ with random effects vector $b_i' = (b_i^{(1)'}, b_i^{(2)'}, \ldots, b_i^{(Q)'})$. The example in [5] included $Q = 22$ outcomes with two random effects each, leading to a 44-dimensional vector b_i of random effects with a 44×44 covariance matrix D. In such high dimensions, computational problems can be avoided by fitting the model to each pair of outcomes respectively. More specifically, the model is fitted to each of the $Q(Q-1)/2$ pairs $\{Y_i^{(1)}, Y_i^{(2)}\}$, $\{Y_i^{(1)}, Y_i^{(3)}\}, \ldots, \{Y_i^{(1)}, Y_i^{(Q)}\}, \{Y_i^{(2)}, Y_i^{(3)}\}, \ldots, \{Y_i^{(Q-1)}, Y_i^{(Q)}\}$. Note that this way of splitting up the data set leads to overlapping sub-samples, as graphically illustrated in panel (d) of Fig. 2.6. Denoting the parameters in pair $\{Y_i^{(p)}, Y_i^{(q)}\}$ by $\theta_{p,q}$, fitting the models to all pairs independently is equivalent to maximizing

$$p\ell(\psi) = \sum_{p<q} \sum_i \ell(Y_i^{(p)}, Y_i^{(q)} | \theta_{p,q}) \qquad (2.12)$$

with respect to $\psi = \{\theta_{1,2}, \ldots, \theta_{Q-1,Q}\}$. Since (2.12) is again a special case of (2.9), inference for ψ immediately follows from pseudo-likelihood theory. As in the previous case the parameter vectors $\theta_{p,q}$ only contain some of the parameters in θ, more specifically only those parameters that appear in the joint model of the pair $\{Y_i^{(p)}, Y_i^{(q)}\}$. Fieuws and Verbeke [5] suggested averaging estimators for the same parameters but other summaries can be explored as well.

5 Concluding Remarks

In this chapter, a general introduction has been given to mixed models. Linear as well as generalized linear mixed models have been discussed. Mixed models currently provide the most flexible tool for the analysis of hierarchical data. Balanced as well as unbalanced data can easily be handled, and associations between observations from the same cluster are modeled through random effects. However, while mixed models can be interpreted as natural extensions of standard linear or generalized linear models to account for clustering, our illustrations in Sect. 3 have shown that parameter interpretation needs careful reflection, that inference is not always following classical asymptotic theory, and that model assessment may be more involved than in models for cross-sectional data. Also, when large data sets are to be analysed, computational difficulties may arise. In many applications however, simplifications can be obtained by splitting up the data such that simpler models can be fitted, the results of which can be combined afterwards, with inference following from pseudo-likelihood theory. The focus of this chapter was on model formulation, parameter interpretation, misconceptions and problems often encountered in practice. Details about various estimation methods, inferential procedures, model selection, and model diagnostics have not been discussed, but can be found in various text books on mixed models and models for clustered data analysis, see, e.g., [4, 15, 19, 30], amongst many others. Many other related topics have not been discussed either in this chapter, including missing data issues [14, 15, 30], Bayesian approaches [10], and nonparametric models for clustered data [33].

Acknowledgements We acknowledge financial support from IAP research Network P7/06 of the Belgian Government (Belgian Science Policy). We are also grateful to Prof. Dr. J. Heyrman of the Department of Public Health and Primary Care of Katholieke Universiteit Leuven (Belgium) to provide us with the data of the Leuven Diabetes Project.

References

1. Arnold BC, Strauss D (1991) Pseudolikelihood estimation: some examples. Sankhya: Indian J Stat Ser B 53:233–243
2. Borgermans L, Goderis G, Van Den Broeke C, Mathieu C, Aertgeerts B, Verbeke G, Carbonez A, Ivanova A, Grol R, Heyrman J (2008) A cluster randomized trial to improve adherence to evidence-based guidelines on diabetes and reduce clinical inertia in primary care physicians in Belgium: study protocol [NTR 1369]. Implement Sci 3:42. https://doi.org/10.1186/1748-5908-3-42
3. Borgermans L, Goderis G, Van Den Broeke C, Verbeke G, Carbonez A, Ivanova A, Mathieu C, Aertgeerts B, Heyrman J, Grol R (2009) Interdisciplinary diabetes care teams operating on the interface between primary and specialty care are associated with improved outcomes of care: findings from the Leuven Diabetes Project. BMC Health Serv Res 9:179. https://doi.org/10.1186/1748-5908-3-42
4. Diggle PJ, Heagerty P, Liang KY, Zeger SL (2002) Analysis of longitudinal data. Clarendon Press, Oxford
5. Fieuws S, Verbeke G (2006) Pairwise fitting of mixed models for the joint modelling of multivariate longitudinal profiles. Biometrics 62(2):424–431
6. Fieuws F, Verbeke G (2009) Joint models for high-dimensional longitudinal data. In: Fitzmaurice G, Davidian M, Verbeke G, Molenberghs G (eds) Longitudinal data analysis. Handbooks of modern statistical methods (Chap. 16). Chapman & Hall/CRC, New York, pp 367–391
7. Fieuws S, Verbeke G, Boen F, Delecluse C (2006) High-dimensional multivariate mixed models for binary questionnaire data. Appl Stat 55(4):1–12
8. Fieuws S, Verbeke G, Molenberghs G (2007) Random-effects models for multivariate repeated measures. Stat Methods Med Res 16(4):387–398
9. Fieuws S, Verbeke G, Maes B, Van Renterghem Y (2008) Predicting renal graft failure using multivariate longitudinal profiles. Biostatistics 9:419–431
10. Gelman A, Carlin JB, Stern HS, Rubin DB (1995) Bayesian data analysis. Texts in statistical science. Chapman & Hall, New York
11. Goderis G, Borgermans L, Grol R, Van Den Broeke C, Boland B. Verbeke G, Carbonez A, Mathieu C, Heyrman J (2010) Start improving the quality of care for people with type 2 diabetes through a general practice support program: a cluster randomized trial. Diabetes Res Clin Pract 88:56–64
12. Goldstein H (1979) The design and analysis of longitudinal studies. Their role in the measurement of change. Academic Press, London

13. McCullagh P, Nelder JA (1989) Generalized linear models, 2nd edn. Chapman & Hall, New York
14. Molenberghs G, Kenward MG (2007) Missing data in clinical studies. Wiley, New York
15. Molenberghs G, Verbeke G (2005) Models for discrete longitudinal data. Springer series in statistics. Springer, New York
16. Molenberghs G, Verbeke G (2007) Likelihood ratio, score, and Wald tests in a constrained parameter space. Am Stat 61:22–27
17. Molenberghs G, Verbeke G, Iddi S (2011) Pseudo-likelihood methodology for partitioned large and complex samples. Stat Probab Lett 81:892–901
18. Pinheiro JC, Bates DM (1995) Approximations to the log-likelihood function in the nonlinear mixed-effects model. J Comput Graph Stat 4:12–35
19. Pinheiro JC, Bates DM (2000) Mixed effects models in S and S-Plus. Springer, New York
20. Ramsay JO, Silverman BW (2005) Functional data analysis. Springer series in statistics, 2nd edn. Springer, New York
21. Raubertas RF, Lee CIC, Nordheim EV (1986) Hypothesis tests for normal means constrained by linear inequalities. Commun Stat Theory Methods 15:2809–2833
22. Shapiro A (1988) Towards a unified theory of inequality constrained testing in multivariate analysis. Int Stat Rev 56:49–62
23. Shock NW, Greullich RC, Andres R, Arenberg D, Costa PT, Lakatta EG, Tobin JD (1984) Normal human aging: the Baltimore longitudinal study of aging. National Institutes of Health Publication 84-2450
24. Silvapulle MJ, Silvapulle P (1995) A score test against one-sided alternatives. J Am Stat Assoc 90:342–349
25. Stram DA, Lee JW (1994) Variance components testing in the longitudinal mixed effects model. Biometrics 50:1171–1177
26. Stram DA, Lee JW (1995) Correction to: variance components testing in the longitudinal mixed effects model. Biometrics 51:1196
27. Thilakarathne PJ, Clement L, Lin D, Shkedy Z, Kasim A, Talloen W, Versele M, Verbeke G (2011) The use of semi-parametric mixed models to analyze PamChip peptide array data: an application to an oncology experiment. Bioinformatics 27:2859–2865
28. Verbeke G, Lesaffre E (1996) A linear mixed-effects model with heterogeneity in the random-effects population. J Am Stat Assoc 91:217–221
29. Verbeke G, Molenberghs G (1997) Linear mixed models in practice : a SAS-oriented approach. Lecture notes in statistics, Number 126. Springer, New York
30. Verbeke G, Molenberghs G (2000) Linear mixed models for longitudinal data. Springer series in statistics. Springer, New York
31. Verbeke G, Molenberghs G (2003) The use of score tests for inference on variance components. Biometrics 59:254–262
32. Verbeke G, Molenberghs G (2007) What can go wrong with the score test ? Am Stat 61:289–290
33. Wu H, Zhang JT (2006) Nonparametric regression methods for longitudinal data analysis. Wiley, New York

Chapter 3
The L2 Impact on Learning L3 Dutch: The L2 Distance Effect

Job Schepens, Frans van der Slik, and Roeland van Hout

Abstract Cross-classified random effect models (CCREMs) are often used for partitioning variation in both experimental and observational linguistic data. However, crossed random effects may have more complex interrelationships than is generally assumed. This becomes clear when comparing first language (L1) and second language (L2) influences on proficiency in Dutch as a third language (L3). Using a large database of L3 speaking proficiency scores, we assessed the mutual dependency between the crossed random effects of the L1 and the L2. The results suggest independent and robust linguistic distance effects of the L1 and the L2: the smaller the linguistic distance to the L3, the higher the L3 proficiency, with the L2 effect being weaker than the L1 effect. Although a model that incorporates an additional L1-by-L2 random interaction effect fits the data best, this model still stipulates the relative importance of an independent L2 distance effect. We found that the L1 distance effect is robust against the L2 distance effect and that the L2 distance effect is robust against interactive effects. We discuss possible explanations for interactions between the L1 and the L2. Overall, the data support independent linguistic distance effects of both the L1 and the L2, besides L1–L2 interactions. We recommend that researchers compare the fit of their crossed random effects models with the fit of models that also include the respective interaction effects.

1 Introduction

Cross-classified random effect models (CCREMs; [1]) are becoming the standard for analyzing linguistic data [2]. A 2008 paper that introduced CCREMs for psycholinguistic study under the heading mixed-effects models [3] had, by September

---

J. Schepens (✉)
Biological Psychology and Cognitive Neuroscience, Freie Universität, Berlin, Germany
e-mail: jobschepens@gmail.com

F. van der Slik · R. van Hout
Centre for Language Studies, Radboud University Nijmegen, Nijmegen, The Netherlands

2015, been cited 2553 times (according to Google Scholar). CCREMs have been used, for example, in studies of linguistic variation [4], syntactic variation in language production [5], and cognate effects in bilingual word recognition [6]. With the exception of some school effectiveness studies, however, few studies have considered the possible interrelatedness between random effects [1, 7–9]. In the present observational study, we examine the consequences of interrelated random effects by modeling the effectiveness of different language backgrounds on proficiency in Dutch as an L3, similar to school effectiveness studies (e.g., [10]).

Being able to use an additional language is widely regarded as helpful for economic mobility and successful integration and in foreign language environments. About half the citizens of the European Union (EU) member states are able to hold a conversation in at least one additional language [11–13]. To what degree do a first (L1) and second language (L2) affect learning an L3? The roles of the L1 and the L2 in L3 perception and production have frequently been addressed in psycholinguistic experiments which are, however, typically characterized by low participant numbers and low numbers of L1s and L2s [14–18]. In addition, participants are often asked to self-report their level of proficiency in an additional language. One way to overcome the limitations of low participant numbers and subjective language proficiency judgments is to use the language testing scores available for large numbers of candidates from state examination databases [19]. Until recently, only a few studies have made use of such data, for example, for automatic error detection [20, 21], but also for assessing the effects of linguistic distance [22]. Experimental evidence shows that the L3 is jointly influenced by both a naturally acquired L1 and an educationally learned L2 [14]. However, L1–L2 interrelatedness has not been investigated on a large scale using CCREMs. The primary aim of this study is to enhance the understanding of how L1s and L2s affect proficiency in an L3. Because the structure of this problem is similar to that of other problems in linguistics and other areas, the approach taken may also be useful for researchers using CCREMs to model experimental and observational data with a complex structure of interrelated random effects.

The degree to which language background influences acquisition of an additional language (cross-linguistic influence) has been "wreathed in controversy" since the emergence of second language acquisition research [23]. Language testing institutions are currently making available large databases containing language testing scores for many learners of an additional language, providing unprecedented opportunities to study complex interrelationships between L1s, L2s, and an L3. In a previous study with CCREMs [24], we found that linguistic distance between the L1 and Dutch as the target language has a substantial and systematic impact on proficiency in Dutch, even with control for variables related to the individual learner and characteristics of the country of origin. Linguistic distance was modeled using measures of the degree of evolutionary change between languages [25, 26], which restricted our analysis to Indo-European languages. Here, we extend this model by testing whether the "best additional language" (i.e., the L2 the participant knows best) has an independent effect of its own and whether such an effect can be explained by linguistic distance, in addition to the patterns previously observed across L1s. We also extend the analysis to non-Indo-European languages, although this means that we cannot apply distance measures based on the degree of evolutionary change. Instead, we gauge linguistic distance in terms of whether the languages belong to a family other than Indo-European (Sino-Tibetan, Niger-Congo, Afro-Asiatic, etc.) and to a genus other than Germanic (Romance, Slavic, Indo-Iranian, etc.).

2 Background

2.1 CCREMs with Interrelated Random Effects

Cross-classified random effect models (CCREMs) are multilevel regression models with crossed random effects that are not completely contained within one another [27, 28]. For example, English is a common second language (L2) for native speakers of languages such as German and Spanish. However, native speakers of these languages may also speak other second languages beside or instead of English. When investigating L3 Dutch proficiency scores across a large number of speakers with many different language backgrounds, one can assume (as we do in this study) that the L1s and L2s involved are drawn randomly from a larger set of languages, ideally from the distribution of all the world's languages. Hence, their effects are called random effects. These random effects are categorical variables, the levels of which are not fixed but randomly sampled. Consequently, if we treat both the L1s and the L2s of the candidates as independently crossed random effects, we assume that the variations in proficiency across L1s and L2s follow independent normal distributions. However, as this chapter will show, real world data sets are often more complicated. One under-investigated issue is that the variation across the levels of a random effect can depend on the variation across the levels of another random effect. For example, although L2 English is common for both L1 German and L1 Spanish, the variation in L3 Dutch proficiency scores due to L2 English is not constant across L1s. For example, L1 Spanish learners may benefit more from L2 English than L1 German learners do. It is clear that the impact of an L2 on learning an L3 cannot be studied without taking into account the impact of the learner's L1.

To fit a CCREM to data, estimation procedures make use of the assumption that multiple random effects vary independently from each other. A fitted CCREM can then be used to find out what part of the variance in a dependent variable relates to each of the random effects (called variance components), and to what extent each of the levels of each random effect contributes to each variance component (called best linear unbiased predictors; BLUPS) [29]. In a CCREM, a response score is predicted by fixed coefficients, random effects, and residuals [7, 30]. Random effects in CCREMs can often be safely assumed to be independent, for example by experimental design [3]. However, the consequences of assuming that random effects are completely mutually independent, for example in observational studies, are not well understood [8, 31]. One way of investigating interrelatedness between random effects is to incorporate an x-by-y random interaction effect, where x and y are the crossed random effects [1]. However, the amount of data available may be insufficient to reliably estimate every x-by-y combination in which the random effects are only "partially balanced" [32] or "partially crossed" [33]. Consequently, many studies avoid taking x-by-y random interaction effects into account.

2.2 Interrelated L1 and L2 Effects

In this study, we compare the available evidence for an independent L2 effect with available evidence for an interactive L1–L2 effect, where the L2 depends on the L1 via an L1-by-L2 random interaction effect. There are various reasons to predict differences between an L1 effect and an L2 effect. Most importantly, L2 learning problems are more commonplace than L1 learning problems. Although virtually all adult L1 learners reach native proficiency levels, many L2 learners struggle to learn even the basics of a foreign language. Generally, having a command of a second language is considered beneficial for further successive language learning. We therefore first hypothesize that the L2 plays a role beyond that of the L1 in explaining L3 proficiency and that this L2 effect also takes the form of a distance effect [18, 34], similar to the L1 distance effect [24]. Furthermore, we hypothesize that L1 and L2 distance effects are generally independent and additive rather than interactive, which would mean that the combination of acquired languages explains L3 proficiency better than either the L1 or the L2. Third, because the L1 is learned earlier, we hypothesize that the L2 effect is less important than the L1 effect in explaining differences in L3 Dutch proficiency. We thus address four questions concerning differences in L3 Dutch proficiency scores across L2s:

1. Is there an L2 effect on L3 Dutch proficiency scores?
2. If so,

    a. Is the L2 effect an additive, independent effect or does it need to be explained in combination with the L1s involved?
    b. Is the L2 effect more or less important than the L1 effect in explaining differences in L3 Dutch proficiency?
    c. Does the L2 effect follow a pattern that is consistent with linguistic distance?

3 Methods

We fit CCREMs on speaking proficiency test scores from the State Examination in Dutch as a Second Language. This examination is developed by the Central Institute for Test Development (Cito) and the Bureau of Inter-Cultural Evaluation (Bureau ICE), two large testing and assessment companies in The Netherlands. The State Examination is a requirement for non-native speakers who want to enroll in a Dutch university; it is also taken by many people who move to the Netherlands for work or marriage. A pass in the speaking part of the exam confirms a B2 level of Dutch on the Common European Framework of Reference for Languages [35, 36]— equivalent to an International English Testing System (IELTS) score of 5.5.

The speaking exam consists of 14 tasks that have to be completed in 30 min. Participants have to provide information, give instructions, give their opinion, and so on. Professional examiners evaluate the content and correctness of the language produced according to formalized testing criteria. The participants can voluntarily fill in a brief questionnaire about various background characteristics. We used the data they provided on the length of their residence in The Netherlands, age at arrival, gender, years of full-time education, country of birth, mother tongue, and best additional language. Best additional language represents the answer to the question: "If you speak another language besides your mother tongue, which other language do you speak? If you speak more than one other language, name the language that you know best."

We drew on a sample of L3 speaking proficiency scores collected from 1995 to 2010. Specifically, we used the first speaking proficiency score recorded for 50,500 unique participants, as some participants attempt the exam multiple times. We included only L1s, L2s, and countries of birth with at least 15 available scores, resulting in a sample with enough data to test learning differences across 73 L1s (median 204.0 speakers per L1), 43 L2s and monolinguals (median 57.5 speakers per L2, including monolinguals), and 122 countries of birth (median 128 speakers). Of the 3212 possible L1–L2 combinations, 759 were observed in the data (216 combinations had at least 15 participants); see Table 3.1 for the 10 most common L1–L2 combinations. 35.7% of all participants had an L2 other than the most common L2 for a particular L1, illustrating the cross-classified nature of the data (the most common L2 was not always English and depended on the L1 of the participants). When English as an L2 was excluded, the data were only slightly more cross-classified (39.2% had an L2 other than the most common L2). Candidates with missing answers on the questionnaire were excluded from the analysis. Candidates with outlying speaking proficiency scores were also excluded. The speaking scores were normally distributed, as shown in Fig. 3.1.

Our previous study showed that speakers of languages closely related to Dutch score higher than speakers of less closely related languages and that educational quality in the country of birth also plays a role [24]. For example, although both come from Switzerland, Swiss native speakers of German on average performed better than Swiss native speakers of French. Furthermore, Spanish native speakers

Table 3.1 The 10 most common L1–L2 combinations and monolingual L1s (positions 7 and 9) by number of speakers

Rank	L1	L2	N
1	German	English	4336
2	Arabic	French	2933
3	Russian	English	2439
4	Arabic	English	2036
5	Spanish	English	1976
6	Polish	English	1733
7	English	–	1666
8	Persian	English	1646
9	Turkish	–	1436
10	Serbian	English	1174

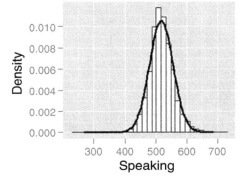

Fig. 3.1 Distribution of speaking proficiency scores

of Spanish on average performed better than Peruvian native speakers of Spanish. Further exploratory analyses showed that bilinguals generally outperformed monolinguals. For example, US citizens who speak L1 English and L2 German on average performed better than US citizens who speak English only. Moroccans who speak L2 French on average performed better than Moroccans who speak Arabic only. And Russians, Iranians, Afghans, etc., who speak English on average outperformed their monolingual counterparts.

We conducted CCREM analysis to investigate differences in L3 Dutch proficiency across L1s, L2s, and L1–L2 combinations more generally. We first estimated a CCREM without incorporating the L2 at all, in which L1s were crossed with countries of birth (C). In other words, we used the model estimated in Schepens et al. [24], but without the binary indicator of L2 presence. This model included the fixed effects of gender, age at arrival, length of residence, years of full-time education, educational quality in the country of birth based on secondary school enrollment rates [37], and an interaction between the latter two covariates. Subsequently, models with more complex random effect structures were fitted to the data and compared using likelihood ratio tests based on -2 logarithms of the likelihood ($-2LL$) under a χ^2 distribution, which can be interpreted as measures of model fit: the probability of observing the data given the maximum likelihood estimates for the model.

The random effect structures were as follows. First, we tested for dependency between the mother tongue (L1) and any additional language (L2), which together constitutes the learners' language background (L1–L2). In this test, we assessed whether a model with one homogeneous random effect of L1–L2 fitted the data better or worse than a model with independently crossed random effects of L1 and L2 (Models 1 and 2). The model with independent effects (Model 2) assumes that the effect of the additional language is constant and irrespective of the learner's mother tongue. The model with homogeneous groups (Model 1) assumes that the effect of the additional language is variable and fully intertwined with the learner's mother tongue. For Model 3, we replaced the crossed effect of the L1 and L2 with a crossed effect of the L1 and L1–L2, effectively allowing for an intertwined effect of additional language that is cross-classified with an effect of the L1. For Model 4, we added the crossed effect of the L2 back into the model.

In summary, four models were tested: a model with a random interaction effect between the L1 and the L2 (Model 1), a model with crossed random effects for L1 and L2 (Model 2), a model with crossed random effects for L1 and a random interaction effect between the L1 and the L2 (Model 3), and a model with crossed random effects for the L1 and the L2, and a random interaction effect between the L1 and the L2 (Model 4). To all of these four models, we also added a crossed random effect of country (C). Moreover, all models included the fixed effects described above. The fixed effects were added to separate variance due to language background from any variance due to confounding variables in terms of individual or country characteristics.

4 Results

Several CCREMs were fitted to the data using different random effect structures as described above. The parameters were estimated using the *lme4* package in R [38]: lmer(Speaking~1 + Gender + Age at Arrival + Length of Residence + Years of Full-time Education * Educational Quality + (1|Country) + (1|L1), data). Note that the interaction term includes the separate covariates. We attempted to improve the fit of this model by comparing different ways of modeling the random effect structure. The model with the best fit to the data included random effects for the L1 (1|L1), the L2 (1|L2), and L1-by-L2 random interaction (1|L1-L2). In the following, we present the results of our model comparison and an analysis of the random by-L1, by-L2, and L1-by-L2 adjustments.

4.1 Model Comparison

Table 3.2 presents the estimates of model fit that were used for model comparison. Several CCREMs were compared to explore the effects of varying the random

Table 3.2 Likelihood ratio tests showing significant improvements of fit against the chi-squared distribution, irrespective of the increasing complexity in random effect structure of the CCREMs

Model no.	Random effects	df	AIC	BIC	−2LL	χ^2
0-0	L1, C	4	495,100	495,136	495,092	
0-1	C	9	493,456	493,536	493,438	1654.08
0-2	L1, C	10	492,692	492,781	492,672	765.90
1	L1–L2, C	10	492,214	492,302	492,194	478.33
2	L1, L2, C	11	492,030	492,127	492,008	186.43
3	L1, L1–L2, C	11	492,008	492,105	491,986	21.35
4	L1, L2, L1–L2, C	12	491,914	492,020	491,890	96.21

All comparisons are significant at the 0.0001 level

Table 3.3 Parameter estimates of the variance components for the intercept-only CCREMs, demonstrating that variation across random effects depends on the total random effect structure incorporated in the model

No.	Residual	C	L1	L1–L2	L2
0-0	32.37 (0.40)	12.97 (4.21)	12.40 (4.21)		
0-1	31.90 (0.40)	12.09 (3.10)			
0-2	31.63 (0.39)	8.72 (2.63)	11.71 (3.89)		
1	31.35 (0.39)	8.84 (2.72)		10.84 (1.76)	
2	31.41 (0.39)	8.44 (2.68)	11.36 (3.86)		3.77 (2.45)
3	31.34 (0.39)	8.23 (2.63)	11.18 (3.99)	5.96 (1.48)	
4	31.34 (0.39)	8.30 (2.72)	11.13 (3.98)	3.29 (1.45)	3.82 (2.60)

The standard deviations are restricted maximum likelihood (REML) estimates and the widths of 95% highest posterior density (HPD) intervals (in parentheses) are constructed from model-specific chains of 20,000 Markov Chain Monte Carlo (MCMC) samples each

effect structure. First, three different null models were fitted to evaluate the effect of controlling for confounding variables, the hierarchical structure of the data, and the fit of a model without L2-related parameters, respectively. The overall null model (model 0-0 in Table 3.2) provides a baseline for the subsequent, more complex models. It has four parameters: a fixed intercept, two random intercepts, and a parameter for residual (student level/level 1) variance. The other null models (models 0-1 and 0-2) are baseline models to demonstrate the hierarchical nature of the dataset. Null model 0-1 shows model fit before the inclusion of language-related random effects; null model 0-2 shows how the addition of control variables changes the model fit.

In null model 0-1, the intra-class correlation is 27.5% (as computed using the values in Table 3.3). More complex random effect structures show that the hierarchical structure of the dataset is more complex due to by-L1 and by-L2 variation, leading to 45.9% of (cross-classified) level-2 variation as compared to total variation in model 4. This demonstrates the importance of incorporating distinct classes in subsequent models, at the same time producing dramatic improvement in fit to the data, as indicated by likelihood ratio tests. Null models 0-1 and 0-2 show highly significant improvements of fit relative to the baseline null model 0-0 due to the added control variables (see Table 3.2). The Akaike information criterion (AIC; −2LL plus twice the number of parameters in the model) and the Bayesian information criterion (BIC; −2LL plus the number of parameters multiplied by the natural logarithm of the sample size) show these patterns as well. Controlling for the six control variables explains part of the variation across countries of birth (32.8%) and mother tongues (5.6%) as compared to the baseline null model 0-0. It is not surprising that explained variance across mother tongues lags behind, given that the null models do not include measures of linguistic distance [24]. Table 3.3 further shows the estimates of the variance components as well as (in parentheses) the 95% highest posterior density (HPD) intervals, which quantify confidence in the parameter estimates shown. According to commonly used criteria [7, 10, 39], the widths of the reported HPD intervals offer no reason to remove any of the parameters from any of the baseline or other models. As an additional check, we inspected all parameter estimates for bimodal patterns using density plots and for deviations from normality.

With the remaining four models (models 1–4 in Tables 3.2 and 3.3, we compared several ways of modeling L1–L2 interrelatedness. The first model assumes that each language background is unique and interactive, and that it is not possible to identify by-L1 or by-L2 variance separately. The second model assumes the opposite, namely that it is not possible to identify an L1-by-L2 random interaction effect. It assumes that by-L1 and by-L2 variance is additive and contributes independently to L3 proficiency. The data provide significantly more support for the second model than for the first ($\chi^2(1) = 186.43, p < 0.0001$), see Table 3.2. This result shows that there is an L2 component in the variance across L3 proficiency scores (question 1), and that there is more evidence for an additive effect than for an interactive effect (question 2a). The parameters also show an increase in the proportion of language-to-country variation, suggesting that the gain in model fit can be attributed to the allocation of remaining variance to a combination of by-L1 and by-L2 variance. The third model assumes that a random interaction effect is a better explanation than an L2 effect. This is confirmed by the data ($\chi^2(1) = 21.35, p < 0.0001$). Furthermore, the by-L2 adjustments depend on by-L1 adjustments. However, an even more complex model fits the data best: By allowing both random interaction and by-L2 adjustment, the fourth model assumes that an L1-independent L2 effect still plays a significant role, alongside random interaction effects. The increase in model fit is again highly significant, showing the importance of the L2, independently of the L1. The estimated parameters for the by-L1 and by-L2 variance indicate that a larger proportion of variance can be attributed to L1 factors than to L2 factors (question 2b). Next, after describing the role of control variables, we will assess the role of linguistic distance in by-L1, by-L2, and L1-by-L2 variance (question 2c).

Table 3.4 Parameter estimates for the fixed predictors included in model 4

Fixed effects	Estimate	2.5% HPD	97.5% HPD
0. Intercept (points)	505.02	498.44	511.36
1. Gender (1 = female)	7.39	6.74	8.05
2. Age at arrival (years)	−0.72	−0.77	−0.68
3. Length of residence (years)	0.62	0.55	0.69
4. Full-time education (years)	−0.77	−1.83	0.24
5. Educational quality (% gross)	0.18	0.11	0.25
6. Interaction 4 * 5	0.04	0.02	0.057

All estimates were significant at the 0.0001 level apart from years of full-time education. The HPD intervals were constructed from one 20,000-sample MCMC chain

4.2 Control Variables

Table 3.4 shows the estimated parameters for the fixed part of model 4. Included in the model are 12 parameters, including six fixed control variables and a fixed intercept. We incorporated the control variables to control the estimations of the random effects for individual differences. The control variables were not centered or otherwise normalized. All the effects were highly significant (apart from that of years of full-time education, see Table 3.4). The gender effect indicates that, all other predictors being equal, female participants scored 7.39 points higher on L3 speaking proficiency (see Fig. 3.1 for the scale), see [40]. Also beneficial were an earlier age at arrival, a longer length of residence, higher educational quality, and a combination of full-time education and quality. Collinearity between educational quality and duration ($r = -0.50$), explains why the sign of the effect of years of full-time education is not in the expected direction. Models with random slopes for duration of full-time education were tested but these converged only sporadically. See [41] for further tests of fixed effects with models that include random slopes. Explorations into various estimated variance and covariance structures revealed only small variations in the way the fixed effects estimates as reported in Table 3.4 deviated, and suggested that age at arrival and education-related effects varied and co-varied across the random effects.

4.3 The L2 Distance Effect

We can now isolate the part of the variance in L3 Dutch proficiency scores that is due to differences across L2s. Figure 3.2 presents the contribution of the by-L2 adjustments to predicted proficiency scores. It shows how the model distributes the estimated L2 variance component of model 4 across the unique L2s. The predicted proficiency scores (dots) represent the by-L2 adjustments of model 4. The lengths of the black lines represent the relative benefit of speaking each of the 18 depicted

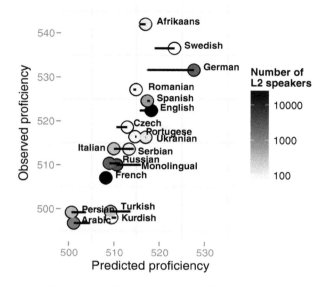

Fig. 3.2 The by-L2 adjustments shift the predicted proficiency scores towards the observed proficiency scores. The dots represent predicted proficiency of model 4 for L2 speakers of the labelled languages. The shading shows the numbers of L2 speakers. The black lines represent the change in predicted (fitted) proficiency between model 4 (including by-L2 adjustments and control variables) and fixed predicted proficiency (including control variables only). Only languages ($N = 18$) with more than 70 L2 speakers are shown

L2s (only the 18 most frequently spoken L2s are shown). The benefit of speaking German is highest (+10.15 points, see also Table 3.6), that of being monolingual is lowest (−5.20 points), and the benefit of Turkish as an L2 is second to lowest (−4.39 points). Figure 3.3 shows the relationship of by-L2 adjustments with by-L1 adjustments ($r = 0.60$, $p < 0.0001$). The graph makes visible a number of interactions; in particular, German is highly beneficial as an L2 for learning Dutch and Italian is only of relatively little benefit. The slope of the plotted regression line suggests that the L2 distance effect is about 1/6 of the size of the L1 distance effect.

We further examined L1 and L2 distance effects by comparing the by-L1 and by-L2 adjustments (BLUPs) for different random effect structures. Table 3.5 orders the top 10 by-L1 adjustments from high to low for the null model with only country and L1 in the random effect structure (model 0-2). The second and third columns show what happens to these estimates when L2 is included in the model. It becomes clear that only slight modifications to the estimated by-L1 adjustments are predicted when L2 variance is accounted for. Two languages in the top 10 switch positions: The L1 benefit of Estonian decreases, whereas that of English increases (both underlined in Table 3.5). It may be the case that the L2s of Estonians (e.g., Russian), when spoken as an L2 by other speakers of other L1s, produce lower L3 proficiency scores. The L2s of native English speakers (e.g., German), on the other hand, may produce

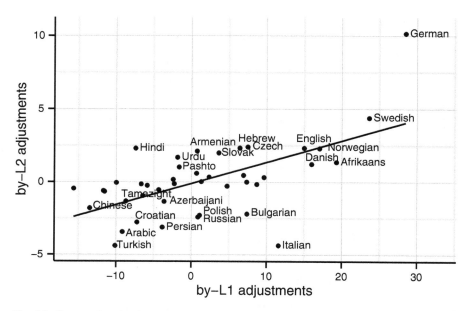

Fig. 3.3 Scatter plot showing the relation between by-L1 and by-L2 adjustments ($r = 0.60$, $p < 0001$). Only L1s for which we have an estimate for the L2 are shown (using model 4). L2s with a by-L2 adjustment higher or lower than 1 point have text labels. The linear regression parameters are $y = 0.15x + -0.11$

Table 3.5 The L1 distance effect is robust against incorporating L2 variance

Language	L1 only	L2 added	Difference
German	25.93	27.07	1.14
Swedish	24.97	24.24	−0.73
Slovenian	21.67	19.97	−1.70
Afrikaans	19.27	19.09	−0.18
Danish	18.96	17.56	−1.40
Norwegian	18.90	17.20	−1.70
Estonian	16.53	14.69	−1.84
Papiamentu	15.08	15.14	0.06
English	12.90	16.29	3.39
Belarusian	12.84	11.55	−1.29

The numbers are aggregated random effects taken from null model 0-2 (L1 only) and model 2 (L2 added). The largest positive and negative differences in the table are underlined

relatively high proficiency scores across other L1s. Table 3.5 shows that the L1 effect is stable across simple and complex models, and particularly for the model that accounts for L2 effects. Furthermore, the higher ranks are populated primarily by linguistically less distant languages—there are five Germanic languages in the

Table 3.6 The L2 distance effect is robust against incorporating L1-by-L2 random interactions

Language	L1 and L2 only	Interactions added	Difference
German	10.11	10.15	0.04
Swedish	4.49	4.39	−0.10
English	3.33	2.32	−1.01
Czech	2.52	2.40	−0.12
Hindi	2.50	2.30	−0.20
Norwegian	2.46	2.30	−0.16
Hebrew	2.23	2.32	0.09
Slovak	1.94	2.00	0.06
Urdu	1.87	1.69	−0.18
Pashto	1.74	1.03	−0.71

The numbers are aggregated random effects taken from model 1 (L1 and L2 only) and model 4 (interactions added)

top six. It may be the case that educational quality in Slovenia was underestimated by our predictors or that—as Slovenian is not widely spoken outside the country—most Slovenians on average speak more than two additional languages besides their mother tongue, which is confirmed in [13].

Column 1 of Table 3.6 shows the top 10 by-L2 adjustments from model 1 (L1 and L2 only). As in Table 3.5, we assess the stability of the L2 distance effect by comparing by-L2 adjustments for model 1 with by-L2 adjustments for model 4 (after the addition of L1-by-L2 random interactions). The ordering is again robust, this time against the addition of L1-by-L2 random interaction effects. The estimate of the adjustment for English changes the most: it decreases by 1.01 points. The L1-by-L2 random interaction may have taken over some part of the adjustment for English—in other words, the variation is decomposed differently, suggesting that the L2 distance effect is not stable for English. One explanation may be that L2 English proficiency is relatively variable relative to L2 proficiency in other languages. Overall, Table 3.6 shows that the L2 effect is robust against interactional effects. There are four Germanic languages in the top six L2s. Although less clear than the L1 distance effect, there seems to be a non-random ordering in the benefits of L2s that, to considerable extent, follows the ordering of an L2 distance effect.

Pairwise comparisons further illustrate the L2 distance effect for specific language backgrounds. On average, bilinguals who speak a closely related language to Dutch as an L2 score higher on speaking proficiency in Dutch than bilinguals with the same mother tongue who speak a less closely related language to Dutch as an L2. This pattern holds for many L1s, with just a few exceptions.

The pairwise comparisons were performed using aggregated random effects to compare the total random variance attributed to each L1–L2 combination. Aggregated random effects were computed by subtracting the fixed predicted proficiency (based only on the fixed predictors) from the fitted scores. We could not use either by-L1 or by-L2 adjustments, as these capture independent variation only. For a number of L2s given an L1, we computed each time whether a specific L2 provided improvement over another by using t-tests between the aggregated random effects at the level of the learners (i.e., computing means, standard deviations, and number of learners for every L1–L2 combination).

The pairwise comparisons showed that bilinguals with L2 English generally performed better than bilinguals with L2 Russian. Pairwise comparisons revealed this pattern in native speakers of Bulgarian ($T = 57.09$, $p < 0.001$), Polish ($T = 114.26$, $p < 0.001$), Lithuanian ($T = 1.62$, $p = 0.108$), Serbian ($T = 62.03$, $p < 0.001$), Pashto ($T = 7.32$, $p = 2.99$), and Armenian ($T = 30.43$, $p < 0.001$); the reverse pattern emerged for native speakers of Persian ($T = -15.30$, $p < 0.001$). The beneficial effect of English as compared to Russian may result from the larger linguistic (and cultural) distance from Russian to Dutch. In addition, bilinguals with L2 English often performed better than bilinguals with L2 French. This pattern was found in native speakers of Polish ($T = 19.99$, $p < 0.001$), Serbian ($T = 15.81$, $p < 0.001$), Russian ($T = 7.03$, $p = 0.001$), and Spanish ($T = 4.76$, $p = 0.001$); the reverse pattern was found for native speakers of Portuguese ($T = -2.71$, $p < 0.01$) and German ($T = -8.51$, $p < 0.001$). Likewise, L2 English was more beneficial than L2 Italian in native speakers of German ($T = 4.87$, $p < 0.001$) and Spanish (T = 16.51, $p < 0.001$) and more beneficial than L2 Spanish in native speakers of German ($T = 5.86$, $p < 0.001$), French ($T = 6.44$, $p < 0.001$), and Portuguese ($T = 7.78$, $p < 0.001$). Bilinguals with L2 German performed even better than bilinguals with L2 English: Pairwise comparisons were significant for native speakers of Czech, French, Polish, Slovak, Russian, Serbian, and Spanish. In addition, bilinguals with L2 German performed better than bilinguals with L2 French, as suggested by the pattern in native speakers of English and Spanish. Because most pairwise comparisons resulted in p-levels below the level of 0.001, we assume that they are robust against the increased possibility of finding a significant effect by chance.

Table 3.7 illustrates the rank ordering of L2s for five L1s; they are mostly in line with predicted L2 distance effects. For example, for Serbian speakers, L2 German is significantly more beneficial than L2 English, which is in turn significantly more beneficial than L2 French, which is significantly more beneficial than L2 Russian, which is significantly more beneficial than no L2 (monolingual). The means displayed in Table 3.7 are all aggregated random effects that can only be interpreted relative to the overall average of the random effects (i.e., L1 Serbian L2 German is 10.21 exam points more beneficial than the overall average adjustment to the fixed predicted score). In all, bilinguals performed better than monolinguals in 29 of 33 pairwise comparisons. Moreover, in 45 of 50 pairwise comparisons, with on average 5 comparisons per language, the pattern emerging was consistent with our hypothesis that distance from Dutch to the additional languages determines proficiency in Dutch. The finding that the pattern was absent in some L1–L2 combinations is consistent with the finding that a model that includes L1-by-L2 random intercepts (Model 4) still provides additional increase in model fit beyond a model that does not include this random interaction (Model 2).

Table 3.7 Pairwise comparisons (*t*-tests) of each estimated group adjustment with its immediately preceding estimated group adjustment (if the L1 is the same)

L1	L2	Estimation, *p* value
Kurdish	English	−1.77
Kurdish	Arabic	−7.92, *p* < 0.0001
Kurdish	Monolingual	−9.28, *p* < 0.0001
Kurdish	Farsi	−13.47, *p* < 0.0001
Kurdish	Turkish	−19.9, *p* < 0.0001
Serbian	German	10.21
Serbian	English	2.89, *p* < 0.0001
Serbian	French	−1.14, *p* < 0.0001
Serbian	Russian	−4.62, *p* < 0.0001
Serbian	Monolingual	−7.89, *p* < 0.0001
Hungarian	German	19.64
Hungarian	Romanian	18.79, *p* < 0.0001
Hungarian	English	16.93, *p* < 0.0001
Hungarian	Monolingual	4.73, *p* < 0.0001
Polish	German	9.44
Polish	English	5.06, *p* < 0.0001
Polish	French	2.53, *p* < 0.0001
Polish	Russian	−0.85, *p* < 0.0001
Polish	Monolingual	−1.88, *p* < 0.0001
Polish	Italian	−3.59, *p* = 0.008
German	French	36.44
German	English	34.12, *p* < 0.0001
German	Italian	31.20, *p* < 0.0001
German	Spanish	31.06, *p* = 0.90
German	Russian	27.67, *p* < 0.0001
German	Monolingual	26.66, *p* = 0.11

The numbers are aggregated random effects taken from model 4 (interactions added). The estimates are statistically controlled for educational differences. The L1s are displayed in no particular order

5 Discussion and Conclusion

By varying the random effect structure of cross-classified random effect models fitted on data from a large number of language learners, we investigated the interrelatedness between the L1 and the L2 in people learning L3 Dutch. The predicted proficiency scores indicate that a significant part of the variation is decomposed into independent L1 and L2 distance effects. A further significant part of the variation is decomposed into an L1-by-L2 random interaction effect. However, comparing by-L2 adjustments for different models shows that the L2 distance effect is robust against interactive models. In addition, pairwise comparisons show that the L2 distance effect is observed repeatedly for different L1s. In the following, we discuss these findings in terms of the concept of linguistic distance and additional language processing.

By-L1 and by-L2 adjustments seem to follow a similar pattern ($r = 0.60$, $p < 0.0001$). As we showed in a previous study [24], linguistic distance between the L1 and Dutch plays a decisive role in learning Dutch as an additional language (75.1% of explained variance). The present findings show that L2 distance also has an effect on learning Dutch as an additional language (answering question 1), although this effect is about six times less strong (question 2b), as shown in Fig. 3.3. The part of the variance modeled by random L1-by-L2 interactions suggests that L1-by-L2 random interactions still play a role. However, a model with independent L1 and L2 components fitted the data significantly better than an interactive model (question 2a).

Our findings suggest that linguistic distance can be measured in terms of proficiency scores in a non-native language. We produced orderings of by-L1 adjustments and by-L2 adjustments that can be used as an empirically validated measure of linguistic distance. Such linguistic distance measures based on additional language proficiency are potentially useful in cross-linguistic influence studies [15, 18] and immigrant studies [9, 42]. Linguists now have the opportunity to take into account not only language classifications into families and genera or phylogenetic distances modeling the degree of evolutionary change between languages [25, 26], but also empirically based measures of linguistic distance.

Linguistically, the finding that the benefit of different L1s and L2s on learning L3 Dutch is not constant raises novel questions for empirical research that may have consequences for the way native and additional language processing is understood. For example, are some languages better suited for non-native language processing than others? Moreover, as the data are consistent with an additive explanation, it seems that the individual types of languages determine learning more than the combination of types.

In sum, we argue that incorporating interactions between random effects into CCREMs challenges the independence that is generally assumed between the different components of a random effect structure. Here, an unbalanced cross-sectional dataset produces considerable support for random interaction effects, which require explanations that are otherwise not considered. For example, in the present case, there may have been differences in the stability of proficiency between L1s and L2s. This possibility is discussed further in [43]. To conclude, interrelated random effects pose challenges to researchers analyzing data with a complex hierarchical structure that have consequences for the interpretation of parameter estimates.

Acknowledgments We thank the participants of the Leuven Statistics Days 2012 for their input and the Secretary of the State Examination in Dutch as a Second Language for providing the data used in this study as well as two anonymous reviewers, Susannah Goss and Theo Bongaerts.

References

1. Raudenbush SW, Bryk AS (2002) Hierarchical linear models: applications and data analysis methods. Sage, Thousand Oaks
2. Barr DJ, Levy R, Scheepers C, Tily HJ (2013) Random effects structure for confirmatory hypothesis testing: keep it maximal. J Mem Lang 68:255–278. https://doi.org/10.1016/j.jml.2012.11.001
3. Baayen RH, Davidson DJ, Bates DM (2008) Mixed-effects modeling with crossed random effects for subjects and items. J Mem Lang 59:390–412. https://doi.org/10.1016/j.jml.2007.12.005
4. Wieling M, Nerbonne J, Baayen RH (2011) Quantitative social dialectology: explaining linguistic variation geographically and socially. PLoS One 6:e23613. https://doi.org/10.1371/journal.pone.0023613
5. Jaeger TF (2008) Categorical data analysis: away from ANOVAs (transformation or not) and towards logit mixed models. J Mem Lang 59:434–446
6. Dijkstra T, Miwa K, Brummelhuis B, Sappelli M, Baayen RH (2010) How cross-language similarity and task demands affect cognate recognition. J Mem Lang 62:284–301
7. Leckie G (2009) The complexity of school and neighbourhood effects and movements of pupils on school differences in models of educational achievement. J R Stat Soc Ser A 172:537–554. https://doi.org/10.1111/j.1467-985X.2008.00577.x
8. Shi Y, Leite W, Algina J (2010) The impact of omitting the interaction between crossed factors in cross-classified random effects modelling. Br J Math Stat Psychol 63:1–15. https://doi.org/10.1348/000711008X398968
9. Van Tubergen F, Kalmijn M (2005) Destination language proficiency in cross national perspective: a study of immigrant groups in nine western countries. Am J Sociol 110:1412–1457. https://doi.org/10.1086/428931
10. Goldstein H, Burgess S, McConnell B (2007) Modelling the effect of pupil mobility on school differences in educational achievement. J R Stat Soc Ser A 170:941–954. https://doi.org/10.1111/j.1467-985X.2007.00491.x
11. European Comission (2001) Special Eurobarometer 55.1: globalization and humanitarian aid. European Commission
12. European Comission (2006) Special Eurobarometer 243: Europeans and their languages. European Commission
13. European Comission (2012) Special Eurobarometer 386: Europeans and their languages. European Commission
14. Bardel C, Falk Y (2007) The role of the second language in third language acquisition: the case of Germanic syntax. Second Lang Res 23:459–484. https://doi.org/10.1177/0267658307080557
15. Cenoz J, Hufeisen B, Jessner U (eds) (2001) Cross-linguistic influence in third language aquisition: psycholinguistic perspectives. Multilingual Matters, Clevedon
16. DeKeyser R (2012) Interactions between individual differences, treatments, and structures in SLA. Lang Learn 62:189–200. https://doi.org/10.1111/j.1467-9922.2012.00712.x
17. Oldenkamp L (2013) The trouble with inflection of adult learners of Dutch: a study on the L1-L2 interplay of morphosyntactic and phonetic-phonological factors. PhD Thesis, Radboud University Nijmegen
18. Ringbom H (2007) Cross-linguistic similarity in foreign language learning. Multilingual Matters, Clevedon
19. Finnie R, Meng R (2005) Literacy and labour market outcomes: self-assessment versus test score measures. Appl Econ 37:1935–1951
20. Nicholls D (2003) The Cambridge Learner Corpus: error coding and analysis for lexicography and ELT. In: Archer D, Rayson P, Wilson A, McEnery AM (eds) Proceedings of the 2003 Corpus Linguistics Conference, pp 572–581

21. Yannakoudakis H, Briscoe T, Medlock B (2011) A new dataset and method for automatically grading ESOL texts. In: Proceedings of the 49th annual meeting of the association for computational linguistics: human language technologies, vol 1. Association for Computational Linguistics, pp 180–189
22. Van der Slik F (2010) Acquisition of Dutch as a second language. Stud Second Lang Acquis 32:401–432. https://doi.org/10.1017/S0272263110000021
23. Kellerman E (1995) Crosslinguistic influence: transfer to nowhere? Annu Rev Appl Linguist 15:125–150. https://doi.org/10.1017/S0267190500002658
24. Schepens J, Van der Slik F, Van Hout R (2013) The effect of linguistic distance across Indo-European mother tongues on learning Dutch as a second language. In: Borin L, Saxena A (eds) Approaches to measuring linguistic differences, Trends in linguistics. Studies and monographs [TiLSM], vol 265. De Gruyter Mouton, Berlin, pp 199–230
25. Bouckaert R, Lemey P, Dunn M, Greenhill SJ, Alekseyenko AV, Drummond AJ, Gray RD, Suchard MA, Atkinson QD (2012) Mapping the origins and expansion of the indo-european language family. Science 337:957–960. https://doi.org/10.1126/science.1219669
26. Holman EW, Brown CH, Wichmann S, Müller A, Velupillai V, Hammarström H, Sauppe S et al (2011) Automated dating of the world's language families based on lexical similarity. Curr Anthropol 52:841–875
27. Beretvas SN (2011) Cross-classified and multiple membership models. In: Hox JJ, Roberts K (eds) The handbook of advanced multilevel analysis. Routledge, New York, pp 313–334
28. Bolker BM, Brooks ME, Clark CJ, Geange SW, Poulsen JR, Stevens MHH, White J-SS (2009) Generalized linear mixed models: a practical guide for ecology and evolution. Trends Ecol Evol 24:127–135. https://doi.org/10.1016/j.tree.2008.10.008
29. Robinson G (1991) That BLUP is a good thing: the estimation of random effects. Stat Sci 6:15–32
30. Rasbash J, Browne WJ (2008) Non-hierarchical multilevel models. In: de Leeuw J, Meijer E (eds) Handbook of multilevel analysis. Springer, New York, pp 301–334
31. Meyers JL, Natasha Beretvas S (2006) The impact of inappropriate modeling of cross-classified data structures. Multivar Behav Res 41:473–497
32. Pinheiro J, Bates D (2000) Mixed-effects models in S and S-PLUS. Springer, New York
33. Raudenbush SW (1993) A crossed random effects model for unbalanced data with applications in cross-sectional and longitudinal research. J Educ Stat 18:321–349. https://doi.org/10.2307/1165158
34. Cenoz J (2001) The effect of linguistic distance, L2 status and age on cross-linguistic influence in third language acquisition. In: Cenoz J, Hufeisen B, Jessner U (eds) Cross-linguistic influence in third language acquisition: psycholinguistic perspectives. Multilingual Matters, Clevedon, pp 8–20
35. Council of Europe (2001) Common European framework of reference for languages: Learning, teaching, assessment. Cambridge University Press, Cambridge
36. Kuijper H, Bergsma A, Bechger T (2004) Staatsexamen NT2. Het gewenste niveau. Deel 1 behoeftepeiling programma II [State Examination in Dutch as a Second Language. The desired level. Part I a needs inquiry for program II]. Central Institute for Test Development, Arnhem
37. UNESCO (2011) The World Bank, Institute for Statistics, School enrollment, secondary (% gross). http://data.worldbank.org/indicator/SE.SEC.ENRR
38. Bates D, Maechler M, Bolker B (2011) lme4: linear mixed-effects models using S4 classes (version R package version 0.999999-0)
39. Baayen RH (2008) Analyzing linguistic data: a practical introduction to statistics using R. Cambridge University Press, Cambridge
40. van der Slik F, van Hout R, Schepens J (2015) The gender gap in second language acquisition: gender differences in the acquisition of Dutch among immigrants from 88 countries with 49 mother tongues. PLoS ONE 10(11):e0142056. https://doi.org/10.1371/journal.pone.0142056
41. Schepens J, Jaeger F, van Hout R (2015) Learning distant sounds: a phonological account of L2 learnability. Utrecht: LOT

42. Chiswick BR, Miller PW (2005) Linguistic distance: a quantitative measure of the distance between English and other languages. J Multiling Multicult Dev 26:1–11. https://doi.org/10.1080/14790710508668395
43. Schepens J, van der Slik F, van Hout R (2016) L1 and L2 distance effects in learning L3 Dutch. Lang Learn 66(1):224–256. https://doi.org/10.1111/lang.12150

Chapter 4
Autocorrelated Errors in Experimental Data in the Language Sciences: Some Solutions Offered by Generalized Additive Mixed Models

R. Harald Baayen, Jacolien van Rij, Cecile de Cat, and Simon Wood

Abstract A problem that tends to be ignored in the statistical analysis of experimental data in the language sciences is that responses often constitute time series, which raises the problem of autocorrelated errors. If the errors indeed show autocorrelational structure, evaluation of the significance of predictors in the model becomes problematic due to potential anti-conservatism of p-values.

1 Introduction

A problem that tends to be ignored in the statistical analysis of experimental data in the language sciences is that responses often constitute time series, which raises the problem of autocorrelated errors. If the errors indeed show autocorrelational structure, evaluation of the significance of predictors in the model becomes problematic due to potential anti-conservatism of p-values.

This paper illustrates two tools offered by Generalized Additive Mixed Models (GAMMs) [10, 19–22] for dealing with autocorrelated errors, as implemented in the current version of the fourth author's MGCV package (1.8.9): the possibility to specify an AR(1) error model for Gaussian models, and the possibility of using factor

---

R. H. Baayen (✉)
Department of Quantitative Linguistics, University of Tübingen, Tübingen, Germany

Department of Linguistics, University of Alberta, Edmonton, AB, Canada
e-mail: harald.baayen@uni-tuebingen.de

J. van Rij
University of Groningen, Groningen, The Netherlands

C. de Cat
University of Leeds, Leeds, UK

S. Wood
University of Bristol, Bristol, UK

smooths for random-effect factors such as subject and item. These factor smooths are set up to have the same smoothing parameters, and are penalized to yield the non-linear equivalent of random intercepts and random slopes in the classical linear framework.

Three examples illustrate the possibilities offered by GAMMs. First, a standard chronometric task, word naming, is examined, using data originally reported in [13]. In this task, and similar tasks such as lexical decision, a participant is asked to respond to stimuli presented sequentially. The resulting sequence of responses constitute a time series in which the response at time t may not be independent from the response at time $t - 1$. For some participants, this non-independence may stretch across 20 or more lags in time. Second, a study investigating the pitch contour realized on English three-constituent compounds [9] is re-examined. As pitch changes relatively slowly and relatively continuously, autocorrelation structure is strongly present. A reanalysis that brings the autocorrelation under statistical control leads to conclusions that differ substantially from those of the original analysis. The third case study follows up on a model reported by DeCat et al. [6, 7] fitted to the amplitude over time of the brain's electrophysiological response to visually presented compound words. We begin with a short general introduction to GAMMs.

2 Generalized Additive Mixed Models

Generalized additive mixed models extend the generalized linear mixed model with a large array of tools for modeling nonlinear dependencies between a response variable and one or more numeric predictors. For nonlinear dependencies involving a single predictor, thin plate regression splines are available. Thin plate regression splines (TPRS) model the response by means of a weighted sum of smooth regular basis functions that are chosen such that they optimally approximate the response, if that response is indeed a smooth function. The basis functions of TPRS have much better mathematical properties compared to basis functions that are simple powers of the predictor (quadratic or higher-order polynomials). Importantly, the smoother is penalized for wiggliness, such that when fitting a GAMM, an optimal balance is found between undersmoothing and oversmoothing.

When a response depends in a nonlinear way on two or more numeric predictors that are on the same scale, TPRS can also be used to fit wiggly regression surfaces or hypersurfaces, approximated by means of weighted sums of regular surfaces which are again penalized for wiggliness. When predictors are not isometric, tensor product smooths should be used. Tensor product smooths (TPS) approximate a wiggly surface or hypersurface using as basis functions restricted cubic splines, again with penalization for wiggliness.

Interactions of numerical predictors with a factorial predictor can be accommodated in two ways. One option is to fit a different wiggly line or surface for each level of such a factor. Alternatively, one may want to take one of the factor levels as reference level, fit a smooth for the reference level, and then fit difference curves or difference surfaces for the remaining factor levels. These difference curves have an interpretation similar to treatment contrasts for dummy coding of factors: The difference curve for level k, when added to the curve for the reference level, results in the actual predicted curve for factor level k.

When a factor has many different levels, as is typically the case for random-effect factors, it may be desirable to require the individual smooths for the different factor levels to have the same smoothing parameter. Together with a heavier penalty for moving away from zero, the resulting 'factor smooths' are the nonlinear equivalent of the combination of random intercepts and random slopes in the linear mixed model.

In what follows, examples are discussed using R, which follows Wilkinson and Rogers [18] for the specification of statistical models. Extensions to the notation for model formulae made within the context of the package for linear mixed models [LME4, 4] and the mgcv package for generalized additive mixed models [19, 20] are explained where used first.

3 Time Series in a Word Naming Task

Although there is awareness in the field of inter-trial dependencies in chronometric behavioral experiments [5, 12, 14, 17], efforts to take such dependencies into account are scarce. De Vaan et al. [8] and Baayen and Milin [2] attempted to take the autocorrelation out of the residual error by including as a covariate the response latency elicited at the preceding trial. This solution, however, although effective, is not optimal from a model-building perspective, as the source of the autocorrelation is not properly separated out from the other factors that co-determine the response latency at the preceding timestep.

To illustrate the phenomenon, consider data from a word naming study on Dutch [13], in which subjects were shown a verb on a computer screen, and were requested to read out loud the corresponding past (or present) tense form. The upper row of panels of Fig. 4.1 presents the autocorrelation function for selected, exemplary, subjects. The autocorrelation function presents, for lags 0, 1, 2, 3, ... the correlation coefficient obtained when the vector of responses v_1 at trials 1, 2, 3, ... is correlated with the vector of responses v_l at trials 1+l, 2+l, 3+l, ... ($l >= 0$). At lag $l = 0$, the correlation is necessarily 1. As the lag increases, the correlation tends to decrease. For some subjects, there is significant autocorrelation at short lags, as illustrated in the first two panels. The subject in the third panel shows a "seasonal" effect, with an initial positive correlation morphing into a negative correlation around lag 10. The subjects in the next two panels show a very different pattern, with autocorrelations persisting across more than 20 lags.

The second row of panels in Fig. 4.1 presents the autocorrelation functions for the residuals of a linear mixed-effects model fitted to the word naming latencies with random intercepts for item (verb) and by-subject random intercepts as well as by-subject random slopes for Trial (the order number of the word in the experimental

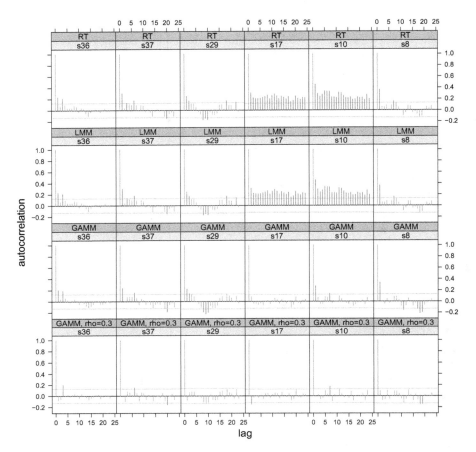

Fig. 4.1 Autocorrelation functions for the residuals of selected participants in the word naming task: top: observed response latencies; second row: residuals of a linear mixed-effects model with random by-participant intercepts and slopes for Trial; third row: residuals of a GAMM with by-participant wiggly curves; fourth row: residuals of a GAMM with by-participant wiggly curves and correction for AR(1) with $\rho = 0.3$

list, i.e., the variable defining the time series in this data set). Using the lme4 package [4] for R (version 3.0.2), the specification of the random effects ((1 + Trial|Subject)) requests by-subject random intercepts, by-subject random slopes for Trial, and a correlation parameter for the random intercepts and slopes.

```
naming.lmer = lmer(RT ~ Regularity + Number + Voicing + InitialNeighbors +
                   InflectionalEntropy + poly(Frequency, 2) + Trial +
                   (1 + Trial|Subject) + (1|Verb),
                   data = naming)
```

Figure 4.1 (second row) indicates that the thick autocorrelational structure for subjects 17 and 10 has been eliminated by the by-subject random regression lines, but the less prominent autocorrelational structure for the other subjects has remained virtually unchanged.

The third row of panels of Fig. 4.1 shows that a GAMM with by-subject factor smooths for Trial, replacing the by-subject straight lines of the linear mixed model yields very similar results. Using the bam function from mgcv for R, the model specification

```
naming.gam = bam(RT ~ Regularity + Number + Voicing + InitialNeighbors +
                InflectionalEntropy + s(Frequency) +
                s(Trial, Subject, bs="fs",m=1) + s(Verb, bs="re"),
                data=naming)
```

requests random intercepts for the verbs (s(Verb, bs="re")) and by-subject wiggly penalized curves for Trial (s(Trial, Subject, bs="fs", m=1), here, bs="fs" requests factor smooths with the same smoothing parameters across subjects, and m=1 requests shrinkage to obtain wiggly random effects).

An improvement is obtained by including an autoregressive AR(1) process for the errors:

$$e_t = \rho e_{t-1} + \epsilon_t, \quad \epsilon_t \sim \mathcal{N}(0, \sigma). \tag{4.1}$$

This equation specifies that the current error is similar to the preceding error by a factor ρ, with Gaussian noise added. As the current error depends only on the preceding error, this is a first-order autoregressive process. Second-order or higher autoregressive process would also take into account the error at $t - k, k = 2, 3, \ldots$. The bam function in the mgcv package offers the possibility of taking a first-order autoregressive process into account by specifying the autoregressive proportionality ρ (with the rho directive in the function call) and by supplying a variable in the data frame, here NewTimeSeries (with levels TRUE, FALSE), indicating the beginning of each new time series with the value TRUE (here, the first trial for each subject), to be supplied to the directive AR.start in the call to bam:

```
naming.r.gam = bam(RT ~ Regularity + Number + Voicing + InitialNeighbors +
                  InflectionalEntropy + s(Frequency) +
                  s(Trial, Subject, bs="fs",m=1) + s(Verb, bs="re"),
                  rho=0.3, AR.start=naming$NewTimeSeries,
                  data=naming)
```

There is no automatic procedure for the selection of the value of ρ. The autocorrelation at lag 1 is a good guide for an initial guesstimate, which may need further adjusting. When changing ρ, it is important not to increase ρ when this does not lead to a visible reduction in autocorrelation, at the cost of inflated goodness of fit and warped effects of key predictors. It should be kept in mind that an AR(1)

Table 4.1 A GAMM fitted to log-transformed picture naming latencies ($\rho = 0.3$); s: thin plate regression spline, fs: factor smooth, re: random effect

A. Parametric coefficients	Estimate	Std. error	t-value	p-value
Intercept	6.5531	0.0512	127.9396	<0.0001
Regularity=regular	0.0093	0.0094	0.9986	0.3180
Number=singular	−0.1147	0.0513	−2.2377	0.0253
Voicing=present	0.0269	0.0101	2.6734	0.0075
Initial Neighborhood Size	0.0179	0.0055	3.2499	0.0012
Inflectional Entropy	−0.0343	0.0159	−2.1616	0.0307
B. Smooth terms	edf	Ref.df	F-value	p-value
s(word frequency)	4.2914	4.6233	7.7445	<0.0001
fs(Trial, subject)	99.4223	358.0000	5.6670	<0.0001
re(verb)	190.1753	280.0000	2.1085	<0.0001

autocorrelative process is only the simplest of possible autocorrelative processes that may be going on in the data, and that hence increasing ρ beyond where it is functional can distort results. The final row of Fig. 4.1 shows that for this example, nearly all autocorrelational structure is eliminated with a small $\rho = 0.3$.

The summary of this model, shown in Table 4.1, shows strong support for the random effects structure for Verb and Subject, with large *t*-values and small *p*-values.[1] Typical examples of by-subject random wiggly curves are shown in Fig. 4.2. These curves capture both changes in intercept, as well as changes over time. For some subjects, the changes are negligible, but for others, they can be substantial, and non-linear.

One could consider replacing the factor smooths by by-subject random intercepts, while at the same time increasing ρ. However, a model such as

```
bam(RT ~ Regularity + Number + Voicing + InitialNeighbors +
    InflectionalEntropy + s(Frequency) +
    s(Subject, bs="re") + s(Verb, bs="re"),
    rho=0.9, AR.start=naming$NewTimeSeries,
    data=naming)
```

provides an inferior fit with an adjusted R-squared of 0.07 (compare 0.36) and an fREML score of 2655 (compare 684). This suggests that in this data set, two very different kinds of processes unfold. One of these processes is autoregressive in

---

[1] The parametric coefficients suggest that regularity is irrelevant as predictor of naming times, that singulars are named faster than plurals, that words with voiced initial segments have longer naming times, as do words with a large number of words at Hamming distance 1 at the initial segment. Words with a greater Shannon entropy calculated over the probability distribution of their inflectional variants elicited shorter response times. A thin plate regression spline for log-transformed word frequency suggests a roughly U-shaped effect (not shown) for this predictor.

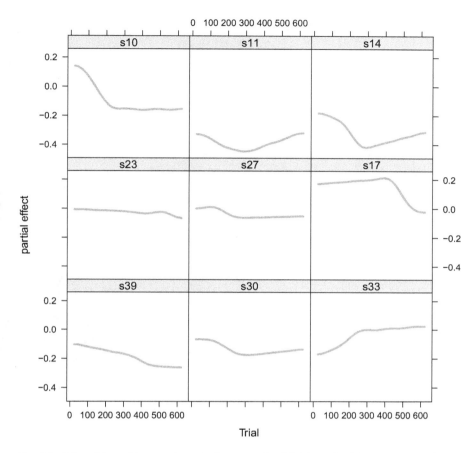

Fig. 4.2 Selected by-subject random wiggly curves for Trial (penalized factor smooths) in the GAMM fitted to word naming latencies

nature, with a relatively small ρ. Possibly, these autoregressive processes reflect minor fluctuations in attention. The other process may reflect higher-order cognitive processes relating to practice and fatigue, such as exemplified by the fastest subject (s11) in Fig. 4.2, who initially improved her speed, but then, as the experiment progressed, was not able to maintain her rapid rate of responding.

Although these task effects typically are not of interest to an investigator's central research question, careful modeling of these task effects is important for the evaluation of one's hypotheses. For instance, the linear mixed effects model mentioned previously does not support an effect of inflectional entropy (Shannon's entropy calculated over the probabilities of a verb's inflectional variants) with $t = -1.87$, whereas the GAMM offers more confidence in this covariate ($t = -2.16$). However, as we shall see next, predictors may also lose significance as autocorrelational structure is brought into the model.

4 Pitch Contours as Time Series

Koesling et al. [9] were interested in the stress patterns of English three-constituent compounds, and measured the fundamental frequency of such compounds as realized by a sample of speakers. In what follows, the response variable of this study, pitch, is measured in semitones.

As can be seen by inspecting the top panels of Fig. 4.3, there are autocorrelations in the pitch contours that are much stronger than those observed for the naming latencies discussed above. In this figure, panels represent the autocorrelation functions for selected *events*, where an event is defined as an elementary time series consisting of the pitch measured at 100 moments in normalized time for the combination of a given compound and a given speaker. Whereas for the naming experiment, there are as many time series as there are subjects, the number of time series in the present phonetics study is equal to the number of unique combinations of subjects and compounds ($12 \times 40 = 480$).

The second row of panels in Fig. 4.3 indicates that a model with by-speaker random intercepts and slopes for (normalized) time does not succeed in consistently reducing the autoregressive structure of this data. Some improvement is achieved when by-subject and by-compound random wiggly curves are added to the model specification (third row of panels), but the errors are only whitened substantially, albeit not completely, by additionally including an autoregressive parameter $\rho = 0.98$ (bottom row of panels). This fourth model was specified as follows.

```
pitch.gam = bam(PitchSemiTone ~ Sex + BranchingOrd +
   s(NormalizedTime) + s(NormalizedTime, by=BranchingOrd) +
   s(NormalizedTime, Speaker, bs="fs", m=1) +
   s(NormalizedTime, Compound, bs="fs", m=1) +
   s(Compound, Sex, bs="re"),
   data=pitch,
   rho=0.98, AR.start=pitch$NewTimeSeries)
```

BranchingOrd is an ordered factor specifying four different compound types (defined by stress position and branching structure). The first smooth, s(NormalizedTime), specifies a wiggly curve for the reference level of this factor. The second smooth term, s(NormalizedTime, by = BranchingOrd), requests difference curves for the remaining three levels of BranchingOrd.[2] A summary of this model is presented in Table 4.2. Figure 4.4 clarifies that the variability across speakers mainly concerns differences in the intercept (height of voice) with variation over time that is quite mild compared to the variability over time present for the compounds.

---

[2]For this to work properly, it is necessary to use treatment contrasts for ordinal factors, in R: options(contrasts = c("contr.treatment", "contr.treatment")).

4 Autocorrelated Errors in Experimental Data in the Language Sciences...

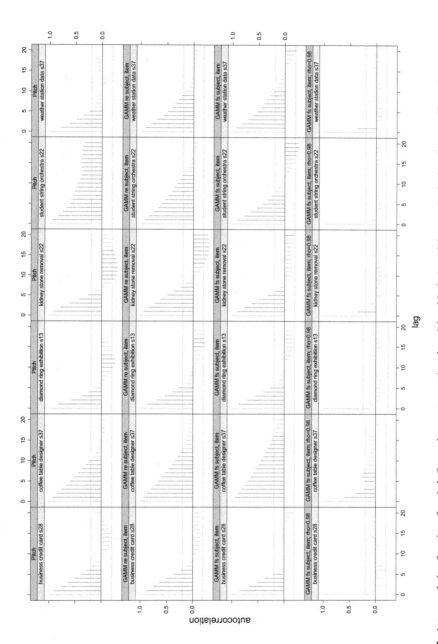

Fig. 4.3 Autocorrelation functions for pitch (in semitones, top row) and model residuals (remaining rows) of selected events. Second row: GAMM with by-participant random intercepts and random slopes for Time and by-compound random intercepts; Third row: GAMM with by-participant and by-compound random intercepts; Fourth row: GAMM with by-compound and by-participant random wiggly curves as well as a correction for AR(1) with $\rho = 0.98$

Table 4.2 Summary of a GAMM for pitch as realized on English three-constituent compounds ($\rho = 0.98$); s: thin plate regression spline, ds: difference spline, fs: factor smooth, re(compound, sex): by-compound random effects for sex

A. Parametric coefficients	Estimate	Std. error	t-value	p-value
Intercept	91.3134	1.4594	62.5689	<0.0001
Sex = male	−13.6336	1.4649	−9.3066	<0.0001
Branching = LN2	0.7739	0.4271	1.8121	0.0700
Branching = RN2	0.2415	0.3657	0.6605	0.5089
Branching = RN3	0.6460	0.4320	1.4955	0.1348
B. Smooth terms	edf	Ref.df	F-value	p-value
s(Time)	7.6892	7.9403	2.7398	0.0064
ds(Time, LN2)	6.5392	7.0804	0.6255	0.7418
ds(Time, RN2)	1.4097	1.5555	2.4744	0.1344
ds(Time, RN3)	6.4987	7.1541	1.9566	0.0411
fs(Time, speaker)	85.7092	105.0000	14.2675	<0.0001
fs(Time, compound)	248.5172	348.0000	3.5294	<0.0001
re(compound, sex)	19.0558	75.0000	0.4566	<0.0001

In principle, one could consider fitting a penalized factor smooth to each of the 480 individual events (time series), although this is currently computationally prohibitively expensive for the large number of events in the present study. The way the model has been specified here is optimistic in the sense that it assumes that how pitch contours are realized can be factored out into orthogonal contributions from individual subjects and from individual compounds. In a more pessimistic scenario, each event makes its own, idiosyncratic, contribution to the model's predictions. In other words, the present model seeks to capture part of the structure in the elementary time series by means of crossed wiggly curves 'by subject' and 'by item'.

Currently, only a single autoregressive parameter ρ can be specified for all events jointly. Inspection of the last row of panels of Fig. 4.4 suggests that it is desirable to relax the assumption that ρ is exactly the same for each event. Although for some events the autocorrelation function is properly flat already for a moderate ρ, see, e.g., the second panel on the first row ($\rho = 0.4$), events remain for which autocorrelations persist across several lags.

Increasing ρ would remove such persistent autocorrelations, but, unfortunately, at the same time induce artificial autocorrelations for other events. This is illustrated in Fig. 4.5, which presents, for four events (rows) the autocorrelation function for increasing values of ρ (columns). For events with hardly any autocorrelation to begin with (upper panels), increasing ρ artificially creates a strong negative autocorrelation at lag 1. The events in the second and third row show how increasing ρ can induce artefactual autocorrelations both at shorter lags (second row) and at longer lags (third row). The event in the fourth row illustrates how increasing ρ attenuates but not removes autocorrelation at shorter lags, while giving rise to new negative autocorrelation at intermediate lags.

4 Autocorrelated Errors in Experimental Data in the Language Sciences...

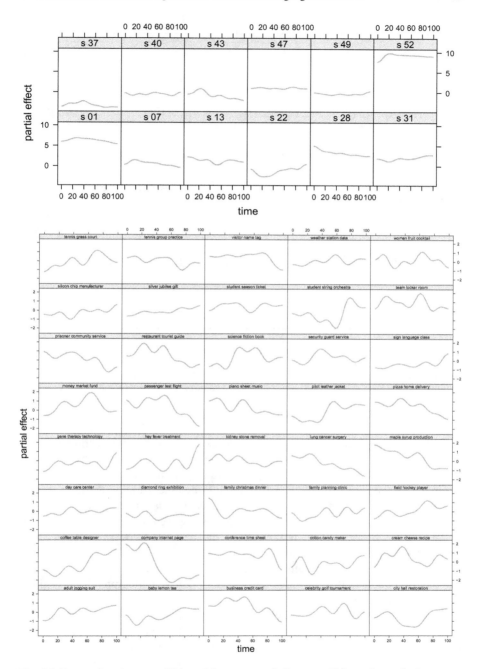

Fig. 4.4 By-speaker (upper trellis) and by-compound (lower trellis) random wiggly curves in normalized time in the GAMM predicting the pitch contour for English three-constituent compounds ($\rho = 0.98$)

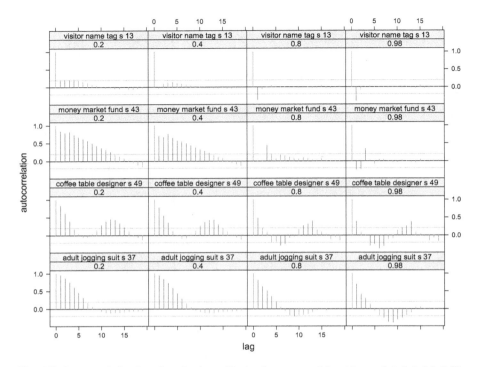

Fig. 4.5 Autocorrelation functions for the residuals of GAMM models with $\rho = 0.2, 0.4, 0.8, 0.98$ (columns) for selected events (rows) where the largest value of ρ, although for most events optimal, induces artificial negative autocorrelations at some lags

Although higher-order autoregressive processes might be more appropriate for many events, they currently resist incorporation into GAMMs. Thus, the analyst is left with two strategies. The first is to select a value of ρ that finds a balance between removing strong autocorrelations, while at the same time avoiding the introduction of artefactual autocorrelation for events which show little autocorrelation to begin with—inappropriate use of ρ may completely obscure the actual patterns in the data.

The second strategy is to remove from the data set those events that show persistent autocorrelations for the optimal ρ obtained with strategy one. When refitting the model to the remaining data points yields qualitatively similar results, it is safe to conclude that the remaining autocorrelational structure in the original model is not an issue.

Two aspects of the present model are of further interest. First, the model includes a thin plate regression smooth for the reference level of compound type (LN1), with difference smooths for the remaining three compound types. Inspection of Table 4.2 reveals only limited support for significant differences between the pitch contours on the four kinds of compounds, and inspection of the difference curves (in panels 2–4 in Fig. 4.6) clarifies that there is little evidence for significant differences with the reference curve. In fact, a simpler model (not shown) with just a spline for normalized time and no main effect or interactions involving branching condition fits the data just as well.

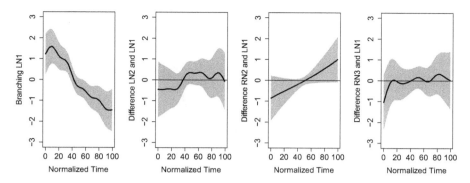

Fig. 4.6 The pitch contour for the LN1 branching condition, and difference curves for the remaining three branching conditions. As the confidence regions for the difference curves always contain the zero line, there is little support for differences in pitch contour as a function of branching condition

The main reason for the absence of the effect of branching condition reported originally by Koesling et al. [9] is the inclusion of the random wiggly curves for compound. When the factor smooth for compound is replaced by random intercepts and random slopes for compound, enforcing linearity, the main effect of branching condition and its interaction with normalized time is fully significant, just as in the original study. This indicates that the variability in the realization of the pitch contours of the individual compounds is too large to support a main effect of branching condition.

We therefore remove branching condition from the model specification, and completing the model with a smooth for the frequency of occurrence of the compound,

```
bam(PitchSemiTone ~ Sex + s(LogFrequency) +
                    s(NormalizedTime) +
                    s(Compound, Sex, bs="re") +
                    s(NormalizedTime, Speaker, bs="fs", m=1) +
                    s(NormalizedTime, Compound, bs="fs", m=1),
    data=pitchc,
    rho=0.98, AR.start=pitchc$NewTimeSeries)
```

we zoom in on the interaction of compound (random-effect factor) by sex (fixed-effect factor), specified above as s(Compound, Sex, bs="re"). Figure 4.7 presents a dotplot for the coefficients for the females on the horizontal axis against the coefficients for the males on the vertical axis. Words for which the males tend to raise their pitch are *passenger test flight, family Christmas dinner*, and *kidney stone removal*, whereas males lower their pitch for *money market fund*. Females, on the other hand, lower their pitch for *tennis grass court, lung cancer surgery*, and *passenger test flight*, but raise their pitch for *maple syrup production, piano sheet music*, and *hay fever treatment*. The two sets of coefficients may even be correlated

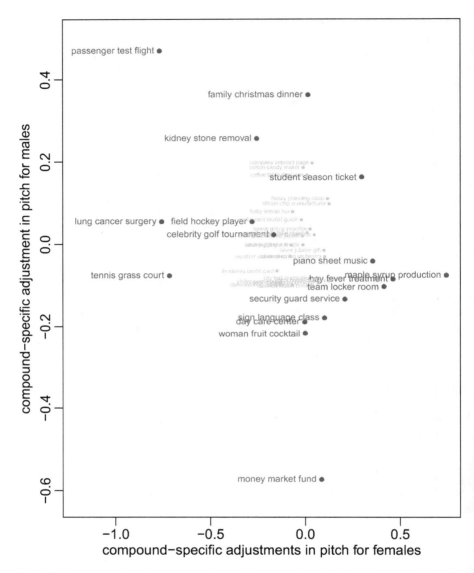

Fig. 4.7 By-compound random contrasts for sex in the GAMM fitted to the pitch contour of English tri-constituent compounds. Positive adjustments indicate a higher pitch

($r = -0.31, t(38) = 0.049$), such that where males substantially raise their pitch, females lower their pitch, and vice versa, possibly reflecting subtle differences in what topics the different sexes find exciting and unexciting (for pitch raising as an index of excitement, see, e.g., [11, 15, 16]).[3]

This case study illustrates three methodological points. First, including random effect curves (by means of factor smooths) for subjects and items may lead to substantially different conclusions about the form of smooth terms in the fixed-effect part of the model specification. Just as including random slopes for a factor X may render the main effect of X non-significant in the context of a linear mixed-effects model, so inclusion of random wiggly curves for a time series t may render an interaction s(t, by=X) non-significant. Second, the coefficients of random-effect interactions such as Compound by Sex may yield novel insights, especially in the presence of correlational structure. Third, when residuals reveal autocorrelational structure, the AR(1) parameter ρ should be chosen high enough to remove substantial autocorrelational structure, but not so high that new, artificial autocorrelational structure is artefactually forced onto the data.

5 Time Series in EEG Registration

Similar to the pitch data, EEG data comprise many small time series, one for each event for which a subject's electrophysiological response to a particular stimulus is recorded. DeCat et al. [6, 7] used English compounds as stimuli, presented in their grammatical order (*coal dust*) and in a manipulated, reversed and ungrammatical order (*dust coal*) to native speakers of English as well as advanced Spanish and German learners of English. The goal of this study was to clarify whether proficiency and language background would be reflected in different electrophysiological processing signatures for these compounds. For the purposes of the present study, the specification of the random-effects structure and the measures taken to bring autocorrelational structure in the residuals under control, and the effects of the choice of ρ on the fixed-effect predictors and covariates in the model are of particular interest. In what follows, the analysis is restricted to the subset of native speakers of English, and to the EEG at channel C1.[4]

---

[3]The details of the coefficients in the present model differ from those obtained in the analysis of Baayen [1]. Thanks to the factor smooths for subject and compound and the inclusion of a thin plate regression spline for word frequency, the present model provides a better fit (AIC 177077.4 versus 187308), suggesting the present reanalysis may provide a more accurate window on sex-specific realizations of compounds' pitch.

[4]Data points with an absolute amplitude exceeding 15 μV, approximately 2.6% of the data points, were removed to obtain an approximately Gaussian response variable.

The model for these data,

```
eeg.gam = bam(Amplitude ~
   s(Time, k=10) + s(Time, by=ConstituentOrder, k=10) +
   te(LogFreqC1, LogFreqC2, k=4) +
   te(LogFreqC1, LogFreqC2, by=ConstituentOrder, k=4) +
   s(LogCompFreq, k=4) + s(LogCompFreq, by=ConstituentOrder, k=4)+
   s(Compound, bs="re")+
   s(Trial, Subject, bs="fs", m=1)+
   s(Time, Subject, bs="fs", m=1),
   data=eegC1, family="scat",
   AR.start=Start, rho=0.85)
```

comprises a smooth for time for the compounds presented with their constituents in the normal order (e.g., *goldfish*), and a difference curve for the condition in which constituent order is reversed (*fishgold*). The model furthermore takes an interaction of the constituent frequencies into account by means of a tensor product smooth, as well as the corresponding difference surface for the reversed order condition. In the light of the very large number of observations (207,600), we slightly lowered the upper bound of the number of basis functions in a given dimension to $k = 4$, in order to avoid fitting overly wiggly surfaces. A thin plate regression spline is introduced to account for the effect of compound frequency, again allowing for a difference between the standard and reversed word order. Random intercepts for compound, and two by-subject factor smooths, one for `Time` and one for the sequence of trials in the experiment (`Trial`, complete the model description. The model summary is given by Table 4.3.

The contributions of the by-subject factor smooths to the model fit is presented in Fig. 4.8. The grey dots represent the by-subject average amplitude for each of

Table 4.3 Generalized additive mixed model fitted to the electrophysiological response of the brain at channel C1 to compound stimuli

A. Parametric coefficients	Estimate	Std. error	t-value	p-value
(Intercept)	0.0552	0.4221	0.1308	0.8960
B. Smooth terms	edf	Ref.df	F-value	p-value
s(Time)	8.5653	8.6645	14.6953	<0.0001
s(Time):Order=reversed	1.5768	1.9624	0.9999	0.4139
s(CompFreq)	1.7242	1.7703	0.7804	0.3172
s(CompFreq):Order=reversed	2.6384	2.8746	21.1108	<0.0001
te(FreqC1,FreqC2)	6.5652	6.6936	4.4840	0.0032
te(FreqC1,FreqC2):Order=reversed	9.6440	10.5906	10.9593	<0.0001
re(Compound)	99.1995	112.0000	10.1991	<0.0001
fs(Trial,Subject)	49.5668	89.0000	12.6940	<0.0001
fs(Time,Subject)	67.4796	89.0000	8.5343	<0.0001

Rev: reversed constituent order in the compound, Norm: normal order. s: thin plate regression spline, te: tensor product smooth, re: random intercepts, fs: factor smooth. ($\rho = 0.85$)

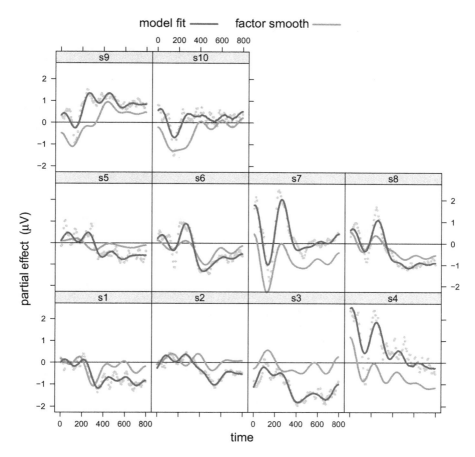

Fig. 4.8 The by-subject factor smooths for Time in the GAMM fitted to the EEG data. Dots represent average response times, the dark gray lines represent the corresponding average for the model fit, and the light gray lines the individual factor smooths

the points in time $t = 4, 8, 12, \ldots$ ms. The dark gray line shows the average of the model fit for the same points in time. The light gray lines visualize the by-subject factor smooths for Trial. Comparing the dark gray and light gray lines, it is clear that a substantial part of the wiggliness of the model fit is contributed by the factor smooths. This figure also illustrates the limitations of the factor smooths: When trends are spiky, as for instance for subjects s5 ad s6 early in time, a strongly penalized smooth will not be able to fit the data points in the spike.

Figure 4.9 illustrates, for four events, that ρ cannot be extended much beyond 0.85 without introducing artefactual negative autocorrelations. Interestingly, changing ρ may have consequences for the predictors of theoretical interest. Figure 4.10 illustrates this point for four smooths in the model. The top panels show that by increasing ρ, the effect of word frequency, which at first blush appears to

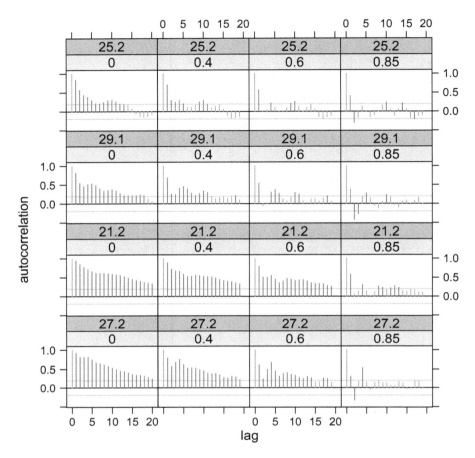

Fig. 4.9 Autocorrelation functions for the residuals of GAMMs fitted to the amplitude of the EEG response to visually presented compounds, for four events (rows), for $\rho = 0, 0.4, 0.6, 0.85$ (columns)

be nonlinear, becomes a straightforward linear effect. The second row of panels clarifies that the difference curve for Time, contrasting the reversed word-order condition with the normal order, is not trustworthy (see also Table 4.3). The increase in the 95% confidence interval that is a consequence of increasing ρ to 0.85, which is required to remove the thick autocorrelative structure in the residuals (Fig. 4.9, left columns), is noteworthy.

The third and fourth rows of Fig. 4.10 illustrate that the regression surface for the frequencies of the compound's constituents depends on constituent order (a threeway interaction of the frequency of the first constituent, the frequency of the second constituent, and constituent order). The contour plots in the third row show the combined effect of the constituent frequencies for the normal constituent order, modeled with a tensor product smooth. Amplitudes are greater along most of the

4 Autocorrelated Errors in Experimental Data in the Language Sciences... 67

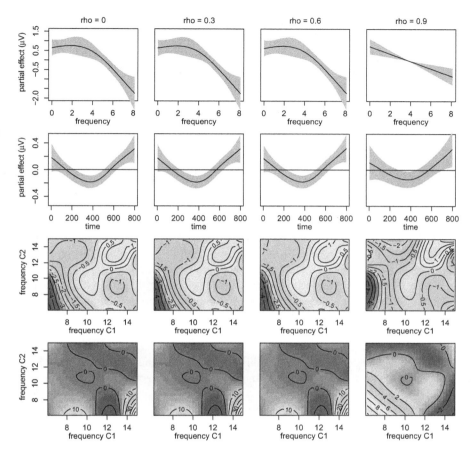

Fig. 4.10 The consequences of increasing ρ from 0 to 0.9 (columns) for the effect of frequency (top), the difference curve for Time contrasting the reversed constituent order with the normal order, the interaction of the frequencies of the first and second constituents (third row), and the difference surface for these predictors contrasting the reversed with the normal constituent order (fourth row)

main diagonal, suggesting qualitative differences in lexical processing for similar versus dissimilar constituent frequencies. For the normal constituent order, this surface is hardly affected by increasing ρ. This does not hold for the corresponding difference surface, as can be seen in the bottom row of Fig. 4.10. In the presence of strong autocorrelations, autocorrelative noise is incorporated into the tensor surface, leading to overaccentuated and uninterpretable patterns in the lower right corner of the partial effect plots. It is only for $\rho = 0.9$ that these irregularities disappear, to give way to a more interpretable difference surface: Amplitudes in the reversed order condition are reduced compared to the normal constituent order when both constituents are of a high frequency, whereas amplitudes increase when both frequencies are low. Thus, this difference surface suggests that the effect of the constituent frequencies in the normal order is largely absent when constituent order is reversed.

In summary, removal of autocorrelative structure in the residuals by means of the ρ parameter for an AR(1) error process may have two important consequences. First of all, analyses will tend to become more conservative. Second, the functional form of nonlinear partial effects may change. In the present examples, excess wiggliness is removed.

6 Concluding Remarks

This study illustrates with three examples the potential of generalized additive mixed models for the analysis of language data: response latencies for reading aloud, pitch contours of three-constituent compounds, and the electrophysiological response of the brain to grammatical and ungrammatical compounds.

GAMMs provide the analyst with two tools for coming to grips with autocorrelational structure in the model residuals: factor smooths and the AR(1) ρ parameter. In the standard linear mixed effects model, systematic changes in how a subject performs over the course of an experiment, or during an experimental trial with a time-series structure, can only be accounted for by means of random intercepts and random slopes. Factor smooths relax this assumption of linearity, and thereby have the potential to provide much tighter fits when random-effect factors indeed behave in a non-linear way.

Autocorrelational structure in the errors may, however, remain even after inclusion of factor smooths. For the reaction times revisited in this study, most of the autocorrelational structure was accounted for by means of factor smooths for the time series constituted by a participant's responses over the time course of the experiment. A mild value of the AR(1) correlation parameter ($\rho = 0.3$) was sufficient to further whiten the residuals. For the pitch data, and the same holds for the EEG data, inclusion of by-participant and by-item factor smooths was not successful at all for removing the autocorrelation. Here, a high value for the AR(1) correlation parameter was necessary for approximate whitening of the errors.

Whitening the errors is important for two reasons (see also[3], for further discussion). First, it protects the analyst against anti-conservative p-values. Second, models with whitened errors are more likely to provide an accurate window on the quantitative structure of the data. The analysis of pitch contours provided an example of the inclusion of a factor smooth rendering a time by fixed-factor interaction non-significant. Furthermore, whitening AR(1) errors may change the functional form of the effect of predictors of interest. The analysis of the EEG data illustrated how an effect that initially seemed nonlinear became straightforwardly linear, as well as a non-linear regression surface that became simplified and better interpretable thanks to whitening.

References

1. Baayen RH (2013) Multivariate statistics. In: Podesva R, Sharma D (eds) Research methods in linguistics. Cambridge University Press, Cambridge, pp 337–372
2. Baayen RH, Milin P (2010) Analyzing reaction times. Int J Psychol Res 3:12–28
3. Baayen R, Vasishth S, Bates D, Kliegl R (2015) Out of the cage of shadows. arxiv.org. http://arxiv.org/abs/1511.03120
4. Bates D, Mächler M, Bolker B, Walker S (2015) Fitting linear mixed-effects models using lme4. J Stat Softw 67(1):1–48
5. Broadbent D (1971) Decision and stress. Academic Press, New York
6. DeCat C, Baayen RH, Klepousniotou E (2014) Electrophysiological correlates of noun-noun compound processing by non-native speakers of English. In: Proceedings of the first workshop on computational approaches to compound analysis (ComAComA 2014). Association for Computational Linguistics and Dublin City University, Dublin, Ireland, pp 41–52
7. DeCat C, Klepousniotou E, Baayen RH (2015) Representational deficit or processing effect? A neuro-psychological study of noun-noun compound processing by very advanced l2 speakers of English. Front Psychol (Lang Sci) 6:77
8. De Vaan L, Schreuder R, Baayen RH (2007) Regular morphologically complex neologisms leave detectable traces in the mental lexicon. Ment Lexicon 2:1–23
9. Koesling K, Kunter G, Baayen RH, Plag I (2012) Prominence in triconstituent compounds: pitch contours and linguistic theory. Lang Speech 56(4):529–554
10. Lin X, Zhang D (1999) Inference in generalized additive mixed models using smoothing splines. J R Stat Soc Ser B 61:381–400
11. Paeschke A, Kienast M, Sendlmeier W (1999) F0-contours in emotional speech. In: Proceedings of the 14th International Congress of Phonetic Sciences, vol 2, pp 929–932
12. Sanders A (1998) Elements of human performance: reaction processes and attention in human skill. Lawrence Erlbaum, Mahwah, NJ
13. Tabak W (2010) Semantics and (ir)regular inflection in morphological processing. PhD thesis, University of Nijmegen. Ponsen & Looijen, Ede
14. Taylor TE, Lupker SJ (2001) Sequential effects in naming: a time-criterion account. J Exp Psychol Learn Mem Cogn 27:117–138.
15. Traunmüller H, Eriksson A (1995) The frequency range of the voice fundamental in the speech of male and female adults. Institutionen för lingvistik, Stockholms Universitet, S-106 91 Stockholm, Sweden
16. Trouvain J, Barry WJ (2000) The prosody of excitement in horse race commentaries. In: ISCA tutorial and research workshop (ITRW) on speech and emotion
17. Welford A (1980) Choice reaction time: basic concepts. In: Welford A (ed) Reaction times. Accademic Press, New York, pp 73–128
18. Wilkinson G, Rogers C (1973) Symbolic description of factorial models for analysis of variance. Appl Stat 22:392–399
19. Wood SN (2006) Generalized additive models. Chapman & Hall/CRC, New York
20. Wood SN (2011) Fast stable restricted maximum likelihood and marginal likelihood estimation of semiparametric generalized linear models. J R Stat Soc (B) 73:3–36
21. Wood SN (2013) On p-values for smooth components of an extended generalized additive model. Biometrika 100:221–228
22. Wood SN (2013) A simple test for random effects in regression models. Biometrika 100:1005–1010

Chapter 5
Border Effects Among Catalan Dialects

Martijn Wieling, Esteve Valls, R. Harald Baayen, and John Nerbonne

Abstract In this study, we investigate which factors influence the linguistic distance of Catalan dialectal pronunciations from standard Catalan. We use pronunciations from three regions where the northwestern variety of the Catalan language is spoken (Catalonia, Aragon and Andorra). In contrast to Aragon, Catalan has an official status in both Catalonia and Andorra, which likely influences standardization. Because we are interested in the potentially large range of differences that standardization might promote, we examine 357 words in Catalan varieties and in particular their pronunciation distances with respect to the standard. In order to be sensitive to differences among the words, we fit a generalized additive mixed-effects regression model to this data. This allows us to examine simultaneously the general (i.e. aggregate) patterns in pronunciation distance and to detect those words that diverge substantially from the general pattern. The results reveal higher pronunciation distances from standard Catalan in Aragon than in the other regions. Furthermore, speakers in Catalonia and Andorra, but not in Aragon, show a clear standardization pattern, with younger speakers having dialectal pronunciations closer to the standard than older speakers. This clearly indicates the presence of a

---

M. Wieling (✉)
Department of Humanities Computing, University of Groningen, Groningen, The Netherlands
e-mail: m.b.wieling@rug.nl

E. Valls
Department of Catalan Philology, University of Barcelona, Barcelona, Spain
e-mail: e.valls@ub.edu

R. H. Baayen
Department of Quantitative Linguistics, University of Tübingen, Tübingen, Germany

Department of Linguistics, University of Alberta, Edmonton, AB, Canada
e-mail: harald.baayen@uni-tuebingen.de

J. Nerbonne
Department of Humanities Computing, University of Groningen, Groningen, The Netherlands

Freiburg Institute for Advanced Studies, University of Freiburg, Freiburg im Breisgau, Germany
e-mail: j.nerbonne@rug.nl

border effect within a single country with respect to word pronunciation distances. Since a great deal of scholarship focuses on single segment changes, we compare our analysis to the analysis of three segment changes that have been discussed in the literature on Catalan. This comparison shows that the pattern observed at the word pronunciation level is supported by two of the three cases examined. As not all individual cases conform to the general pattern, the aggregate approach is necessary to detect global standardization patterns.

1 Introduction

In this study we investigate a Catalan dialect data set in order to identify social and linguistic factors which play an important role in predicting the distance between dialectal pronunciations and the Catalan standard language (which is a formal variety of Catalan mainly based on the dialects of the eastern counties of Catalonia, including those of the Barcelona area). We use Catalan dialect pronunciations of 320 speakers of varying age in 40 places located in three regions where the northwestern variety of the Catalan language is spoken (the autonomous communities Catalonia and Aragon in Spain, and the state of Andorra). Our approach allows us to investigate border effects caused by different policies with respect to the Catalan language. As the Catalan language has been the native and official language (i.e. used in school and in public media) of both Andorra and Catalonia, but not in Aragon,[1] we will contrast these two regions in our analysis.

We show that the speakers of Catalan in Catalonia and Andorra use a variety of Catalan closer to the standard than those in Aragon. Because this tendency is particularly strong among younger speakers, we argue that it is at least in part due to the introduction of Catalan as an official language in the 1980s in Catalonia and Andorra but not in Aragon. Naturally the differences we find may have existed before the language became official in Catalonia, but this cannot explain the larger differences among the young.

Since we suspect that the changes associated with standardization will be far-ranging, we deliberately conduct our analysis in a way that is likely to detect a wide range of differences, effectively aggregating over all differences with respect to the standard in each variety we examine. By taking into account many variables, we deliberately deviate from common sociolinguistic practice which typically focuses on only a small number of variables. We cast a wider net in an effort to obtain a more comprehensive (i.e. aggregate) view, and avoid selecting only those variables

---

[1] In Andorra, Catalan is the only official language. In Catalonia, where Spanish and Aranese (a variety of Occitan) are also official, Catalan was the vehicular language of education during the 1920s and the 1930s and achieved this status again after Franco's dictatorship in the early 1980s [1]. That means that all subjects except second and third languages are taught in Catalan in the public schools of Catalonia and Andorra. In Aragon, Catalan has only been a voluntary subject in schools in the eastern counties (where Catalan is spoken) since 1984 [2]. The standard variety used at all schools in these areas is the one sanctioned by the *Institut d'Estudis Catalans* [3].

that behave as predicted. In a second step, we will investigate whether the aggregate pattern observed at the word pronunciation level also holds when focusing on the more commonly investigated sound (phonemic) level.

1.1 Border Effects

Border effects in European dialectology have been studied intensively (see [4] for an overview). In most of these studies, border effects have been identified on the basis of a qualitative analysis of a sample of linguistic features. In contrast, Goebl [5] used a dialectometric approach and calculated aggregate dialect distances based on a large number of features to show the presence of a clear border effect at the Italian–French and Swiss–Italian borders, but only a minimal effect at the French–Swiss border. This approach is arguably less subjective than current practice in social dialectology (focusing on a pre-selected small set of items), as many features are taken into account simultaneously and the measurements are very explicit. However, Woolhiser [4] is very critical of this study, as Goebl does not discuss the features he used and also does not consider the sociolinguistic dynamics as well as ongoing dialect changes (i.e. he uses static dialect atlas data).

Border effects have generally been studied with respect to national borders. In the present paper, we focus on one language border within a single nation state, and on a second border between two states. The former kind of border has been scarcely studied at all [4].

Several researchers have offered hypotheses about the presence and evolution of border effects in Catalan. For example, Pradilla [6, 7] indicates that the border effect between Catalonia and Valencia might increase, as the two regions recognize different varieties of Catalan as standard (i.e. the unitary Catalan standard in Catalonia and the Valencian Catalan substandard in Valencia). In a similar vein, Bibiloni [8] discusses the increase of the border effect between Catalan dialects spoken on either side of the Spanish–French border in the Pyrenees during the last three centuries. More recently, Valls et al. [9] conducted a dialectometric analysis of Catalan dialects and found, on the basis of aggregate dialect distances (average distances based on hundreds of words), a clear border effect contrasting Aragon with Catalonia and Andorra. This dialectometric approach is an improvement over Goebl's [5] approach, since Valls et al. measure dialect change by including pronunciations for four different age groups (measuring dialect evolution by the apparent-time construct; [10]). However, it ignores other sociolinguistic variables due to its purely dialectometric nature.

1.2 Combining Dialectometry and Social Dialectology

The methodology used in the present study essentially follows dialectometry, which has generally focused on determining aggregate pronunciation distances, and the geographical pattern of aggregate variation ([11], Chap. 1). In contrast, many dialectologists have focused on the influence of specific social factors on the realization of (individual) linguistic variables. Instead of examining a large set of items simultaneously, however, social dialectologists have generally investigated smaller sets of pre-selected linguistic variables.

We grant the essential correctness of Woolhiser's [4] critique that dialectometry has at times been blind to the potential importance of non-geographic conditioning factors. Therefore, in this study, we combine perspectives from two approaches, dialectometry and social dialectology. Following dialectometry, we will measure distances for a large set of dialectal pronunciation data, preventing in this way biased choices in the selection of material [12]. (Of course, as we work with a pre-existing pronunciation data set our analysis will be biased as well towards the material included in this set.) In line with social dialectology, however, in analyzing these distances, we will also take several social factors into account. We have not conducted surveys to determine how the differences we measure are perceived socially. In this sense, we are not in a position to gauge the social meaning of the changes we examine, as sociolinguists often expect. We nonetheless explore the hypothesis that linguistic changes are being brought about by a social change, namely the change to using standard Catalan in schools and public media in part of the Catalan-speaking area. In this sense we are conducting a sociolinguistic study.

In addition, we aim to clarify the relationship between aggregate (dialectometric) analyses, which often ignore the linguistic details most responsible for aggregate relations, and analyses based on selected linguistic features (most non-dialectometric analyses). While dialectometric analyses have aimed at establishing the relations among varieties, analyses based on selected linguistic features such as rhotacization, the raising of front vowels or varying verbal inflections are often motivated both by the wish to establish the social affinities of variation, but also by the wish to adduce linguistic structure in the variation.[2]

1.3 *Hypotheses*

In our analysis we will contrast the area where Catalan is recognized as an official language (Catalonia and Andorra) with the area where it is not (Aragon). This contrast allows us to investigate the influence of an internal border within the same country (i.e. Aragon versus Catalonia) as opposed to a national border (Andorra–Spain). Based on the results of Valls et al. [9], we expect to observe larger pronunciation distances from standard Catalan in Aragon than in the other two

---

[2]Wieling and Nerbonne [13, 14] summarize several earlier attempts to ascertain the linguistic foundations of aggregate dialectometric differences, so we shall not review those here.

regions.[3] More importantly, however, we expect that the models will differ with respect to the importance of the sociolinguistic factors. Mainly, we expect to see a clear effect of speaker age (i.e. with younger speakers having pronunciations closer to standard Catalan) in the area where Catalan has the status of an official language, while we do not expect this for Aragon, as there is no official language policy which might 'attract' the dialect pronunciations to the standard. In contrast to the exploratory visualization-based analysis of Valls et al. [9], our (regression) analysis allows us to assess the significance of these differences. For example, while Valls et al. [9] state that urban communities have pronunciations more similar to standard Catalan than rural communities, this pattern might be non-significant (as they reach this conclusion on the basis of visualization only).

In addition we shall examine a methodological hypothesis, namely that the standardization we are interested in will be more insightfully investigated from an aggregate, dialectometric perspective rather than from the perspective of a small number of sound changes. In defense of the plausibility of this view we note that standardization efforts are unlikely to be undertaken if only a small number of linguistic items is at stake. Standardization normally involves a large number of changes, certainly when viewed from the perspective of all the different varieties affected. However, while we do intend to examine this hypothesis, we do not propose to test it rigorously in this study.

2 Material

2.1 *Pronunciation Data*

The Catalan dialect data set contains basilectal phonetic transcriptions (using the International Phonetic Alphabet) of 357 words in 40 dialectal varieties and the Catalan standard language. The locations are spread out over the state of Andorra (two locations) and two autonomous communities in Spain (Catalonia with 30 locations and Aragon with 8 locations). In all locations, Catalan has traditionally been the dominant language. Figure 5.1 shows the geographical distribution of these locations. The locations were selected from 20 counties, and for each county the (urban) capital as well as a rural village was chosen as a data collection site. In every location eight speakers were interviewed, two per age group (F1: born between 1991 and 1996; F2: born between 1974 and 1982; F3: born between 1946 and 1960; F4: born between 1917 and 1930). All data was transcribed by a single transcriber

---

[3]It might be argued that this pattern is due to the fact that the Catalan standard language is mainly based on the eastern dialects of Catalonia. Although it is true that the northwestern varieties of Catalonia and Andorra have historically converged towards the (closer and more prestigious) eastern varieties during the twentieth century, Valls et al. [9] have shown that the standardization process has been much more effective in the diffusion of the prestigious features westwards.

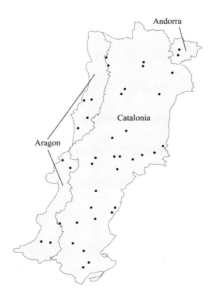

Fig. 5.1 Geographical distribution of the locations. Two locations are found in Andorra, eight in Aragon and the remaining 30 locations are found in Catalonia

(Esteve Valls), who also did the fieldwork for the youngest (F1) age-group between 2008 and 2011. The fieldwork for the other age groups was conducted by another fieldworker (Mar Massanell) between 1995 and 1996. The complete data set we use contains 357 items, consisting of 16 articles, 81 clitic pronouns, eight demonstrative adjectives, two neuter pronouns, two locative adverbs, 220 inflected forms of five verbs, 20 possessive adjectives and eight personal pronouns. The complete item list and a more detailed description of the data set are given by Valls et al. [9]. Note that the data set did not contain any nouns and only contained a limited number of verbs. The fact that over 60% of the words studied are forms of only five verbs means that the sample is biased toward these words. A follow-up study using different material would be worthwhile. However, these five verbs are representative of the five regular paradigms in Catalan and allow us to take into account all the regular inflections of the Catalan verbs.

The standard Catalan pronunciations were transcribed by the second author and are based on the *Gramàtica Catalana* [3] and the proposal of the *Institut d'Estudis Catalans* for an oral Standard Catalan language [15, 16].

2.2 Sociolinguistic Data

Besides the information about the speakers present in the corpus (i.e. gender, age and education level of the speaker), we extracted additional demographic information about each of the 40 locations from the governmental statistics department of Catalonia [17], Aragon [18] and Andorra [19]. The information we extracted for each location was the number of inhabitants (i.e. community size), the average community age, the average community income, and the relative number of tourist beds (i.e. per inhabitant; used to estimate the influence of tourism) in the most recent year available (ranging between 2007 and 2010). There was no location-specific income information available for Andorra, so for these two locations we used the average income of the country [20].

As the data for the older speakers (age groups F2, F3 and F4) was collected in 1995, the large time span between the recordings and measurement of demographic variables might be problematic. We therefore obtained information on the average community age, average community income and community size for most locations in 2000 (which was the oldest data available online). Based on the high correlations between the data from the year 2000 and the most recent data for each of the separate measures (in all cases $r > 0.9$, $p < 0.001$), we decided to use the most recent demographic information in this study. No historical information about the number of tourist beds was available for Catalonia and Aragon, but we do not have reason to believe that this correlation strength should be lower than for the other variables (and thus we can use the most recent data).

3 Methods

3.1 Obtaining Pronunciation Distances

For all 320 speakers, we calculated the pronunciation distance between the standard Catalan pronunciations and their dialectal counterparts by using a modified version of the Levenshtein distance [21]. The Levenshtein distance transforms one string into the other by minimizing the number of insertions, deletions and substitutions. For example, the Levenshtein distance between two Catalan variants of the word 'if I drank', [beɣésa] and [bejɣɛ́s] is 3:

be ɣésa	insert j	1
bejɣésa	subst. ɛ́ for é	1
bejɣɛ́sa	delete a	1
bejɣɛ́s		
		3

This sequence corresponds with the following alignment:

b	e		ɣ	é	s	a
b	e	j	ɣ	ɛ́	s	
		1		1		1

The standard Levenshtein distance does not distinguish vowels from consonants and therefore could align these together. In order to prevent these (linguistically) undesirable alignments, a syllabicity constraint is normally added, allowing only alignments of vowels with vowels, consonants with consonants, and /j/ and /w/ with both consonants and vowels. It prevents alignments of other sounds, as these are assigned a very large (arbitrary) distance [22, 23].

It is clear that these Levenshtein pronunciation distances are very crude as the Levenshtein algorithm does not distinguish (e.g.,) substitutions involving similar sound segments, such as /e/ and /ɛ/, from more different sound segments, such as /e/ and /u/. Wieling et al. [24] proposed a method to automatically obtain more sensitive sound segment distances on the basis of how frequent they align according to the Levenshtein distance algorithm. Sound segments aligning relatively frequently obtain a low distance, while sound segments aligning relatively infrequently are assigned a high distance. The sound distances are based on calculating the Pointwise Mutual Information score (PMI; [25]) for every pair of sound segments. The automatically obtained sound segment distances were found to be phonetically sensible (based on six independent dialect data sets; [26]) and also improved pronunciation alignments when these sound segment distances were integrated in the Levenshtein distance algorithm [24]. A detailed description of the PMI-based approach can be found in Wieling et al. [26]. Similar to the study of Wieling et al. [27] on pronunciation differences between Dutch dialects and standard Dutch, our pronunciation distances are not based on the Levenshtein distance (with syllabicity constraint), but rather on the PMI-based Levenshtein distance. Using this phonetically more sensitive measure, the difference of the example alignment shown above is 0.107. The calculation is illustrated below:

b	e		ɣ	é	s	a
b	e	j	ɣ	ɛ́	s	
		0.0339		0.0345		0.0388

On average, longer words will have a greater pronunciation distance (i.e. more sounds may change) than shorter words. Therefore we normalize the PMI-based word pronunciation distances by dividing by the alignment length. Since the distribution of the Levenshtein distances was skewed, we log-transformed these distances (after adding a small value, 0.01, to prevent taking the log of 0). Note that log-transforming the PMI-based Levenshtein distances has been previously reported to increase the match with perceptual distances (for native-likeness; [28]). After log-transformation, we centered the Levenshtein distances (i.e. subtracted the mean value). Consequently, a Levenshtein distance of 0 indicates the average Levenshtein distance, whereas negative and positive values are indicative of Levenshtein distances lower or higher than the average, respectively.

3.2 Mixed-Effects Regression Modeling

The usefulness of a generalized linear mixed-effects regression model (GLMM) in language variation research has already been argued for extensively by Tagliamonte and Baayen [29]. In summary, a generalized linear mixed-effects regression model allows the researcher to determine which variables (i.e. predictors) are important in language variation, while also taking into account that the interviewed informants as well as the specific linguistic items included are a source of variation. While the GLMM is suitable to determine the preference for a certain form over another (e.g., *was* versus *were* in the study of [29]), the dependent variable may also be numerical instead of binary. In our case, the numerical dependent variable will be the pronunciation distance from standard Catalan on the basis of the log-transformed and centered PMI-based Levenshtein distance.

As explained by Tagliamonte and Baayen [29], a mixed-effects regression model distinguishes fixed-effect factors from random-effect factors. Fixed-effect factors have a small (fixed) number of levels that exhaust all possible levels (e.g., gender is either male or female), while random-effect factors have levels sampled from a large population of possible levels (e.g., we use 357 words, but could have included other words). A mixed-effects regression analysis allows us to take the systematic variability linked to our speakers, locations and words (i.e. our random-effect factors) into account. For example, some words might (generally) be more similar to standard Catalan than other words. By estimating how much more similar these words are, the general regression formula can be adapted for every individual word. These adjustments to the general model's intercept are called 'random intercepts'. For example, Fig. 5.2 shows the effect of the (standardized) year of birth of the speakers on the (log-transformed and centered) linguistic distance from standard Catalan for two different words, *meves* 'my' (feminine plural possessive), and *ell* 'he'. In these graphs, each circle corresponds to the pronunciation of *meves* (left graph) or *ell* (right graph) of a single speaker. The dashed line (which is the same in both graphs) indicates the general effect (across all words) of the year of birth of the speaker on the linguistic distance from standard Catalan (i.e. the fixed effect). It shows a slightly negative slope, with the intercept (i.e. the height at where the standardized year of birth of the speaker equals zero; the reason for standardizing the predictors is explained below) being close to zero. The solid line in each graph shows the word-specific effect of year of birth of the speaker on the linguistic distance from standard Catalan (i.e. the fixed effect plus the random intercept and random slope; see below). Clearly, the solid line belonging to the word *meves* has an intercept which is higher than the dashed line (i.e. *meves* generally has a higher linguistic distance from standard Catalan than the average word), while the solid line of *ell* is positioned much lower (and thus *ell* is, on average, more similar to standard Catalan).

Similarly, the effect of a certain predictor may also vary per word. For example, while in general younger speakers may have pronunciations closer to standard Catalan than older speakers (shown by the dashed line in Fig. 5.2 whose slope is

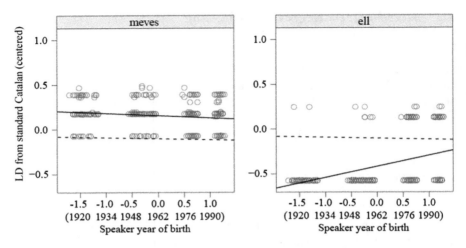

Fig. 5.2 Example of random slopes and intercepts for the standardized year of birth of the speaker per word. For ease of interpretation, the actual year of birth values have been added below the standardized values. The dashed line indicates the general model estimate (the intercept and the coefficient for speaker year of birth) for all words, while the solid lines indicate the estimates of the intercept and the slope for the two words (i.e. the total effect: fixed-effect intercept and slope plus random intercept and slope). The circles represent the distances for individual variants of the words *meves* (left) and *ell* (right). The dependent variable was centered, so an LD of 0 indicates the mean distance from standard Catalan

slightly negative) the precise effect could vary per word. Some words may even show a completely opposite pattern, with older speakers having pronunciations closer to standard Catalan. These (by-word) random slopes, in combination with the random intercepts, allow the regression formula to be adapted for every individual word (or other random-effect factor). For example, the solid lines in Fig. 5.2 show that the effect (i.e. slope) of the year of birth of the speaker for the word *meves* is slightly more negative than the general pattern (i.e. younger speakers use a pronunciation closer to standard Catalan), while the effect for the word *ell* shows the opposite pattern with a positive slope. For the word *ell*, younger speakers have adopted a slightly different pronunciation ([éj]) than the one used in standard Catalan and by older speakers ([éʎ]), as the sound [ʎ] is disappearing from most young phonetic inventories.

In order to prevent type-I errors, it is important to consider both random intercepts as well as random slopes [29–33]. A more detailed introduction about mixed models applied to language data is given by Baayen [34] and Baayen et al. [30]. While Barr et al. [31] advocate an approach where the random-effects structure is maximally complex, we do not favor this approach given the large size of our dataset. Furthermore, Bates et al. [32] show that the approach of Barr et al. [31] may result in overfitting and convergence errors. Consequently, we will only fit the random-effects structure supported by the data.

3.3 Generalized Additive Mixed-Effects Regression Modeling

The difference between a generalized additive model (GAM; [35]) and the generalized linear regression model explained earlier is that the former allows the explicit inclusion of non-linear relationships via so-called smooths. While non-linearities can be included in a generalized linear regression model, in that case the specific form (e.g., a parabola) needs to be specified in advance. A generalized additive mixed-effects regression model does not require a predefined form, but rather determines the shape of the relationship (i.e. modeled by so-called smooths) itself. Furthermore, a smooth can contain multiple numerical variables and thus represent a (potential) non-linear surface. Importantly, if a pattern is linear rather than non-linear, the GAM smooth will reflect this as well. Consequently, it is more flexible than (generalized) linear mixed-effects regression.

There are several choices to make regarding the smooths. First of all, the researcher has to choose the basis functions for each smooth. For example, smooths may consist of a series of cubic polynomials (i.e. a cubic regression spline). Another type of basis function is the thin plate regression spline, which is a combination of several simpler functions (such as a linear function, a quadratic function, a logarithmic function, etc.). Furthermore, a limit needs to be specified for the complexity of each smooth. For a cubic regression spline, this limit is specified as the number of knots, which are the points at which the cubic polynomials are connected. The higher this number, the more cubic polynomials may be used to model the smooth. For the thin plate regression spline, which is the basis function we use (as it is the best approximation of the optimal fit; [36]), the complexity is limited by the number of simpler functions used to model the smooth. The actual complexity of the smooth is indicated by estimated degrees of freedom (edf). If the edf value is equal to 1, the smooth models a linear pattern, whereas an edf value higher than 1 indicates a non-linear pattern. Importantly, visualization is essential to investigate the specific shape of the smooth.

Crucially, overfitting is prevented internally by using cross-validation. Furthermore, the GAM implementation we use (i.e. the *mgcv* R package, version 1.8.8; [37, 38]) allows that random intercepts and slopes are included as well. In this generalized additive modeling framework, random intercepts and slopes are represented by smooths with an associated *p*-value, indicating if their inclusion is necessary or not. Consequently, model comparison is not required to assess if random intercepts and slopes are necessary to include.

An important focus of dialectometry is the relationship between dialect distance and geographic location (e.g., see [39]). While it has become standard practice to analyze the influence of geography on language variation by using geographic distance as an independent variable [40], this approach necessarily assumes that locations having the same distance from some reference point are relatively similar (irrespective of their absolute position). This is obviously not very flexible, and does not allow for distinct, irregularly shaped dialect areas (as the effect of distance is assumed to be the same in every direction). Instead of using distance, we fit a more flexible two-dimensional non-linear surface to the dialect data, with as geographical predictors the longitude and latitude of the locations for which dialect data is available. In this way, geography is modeled by a two-dimensional surface, rather than a set of distances. Of course, the random-effect factor location (i.e. the random intercept for location) would also be able to model the effect of geography (if the geographical smooth were absent). However, such an approach would not take advantage of the fact that people living in nearby locations generally have a more similar pronunciation than those living far apart.

Instead of using a generalized linear mixed-effects regression model, we therefore use a generalized additive mixed-effects regression model where geography is modeled by a non-linear interaction (represented by a two-dimensional thin plate regression spline) of longitude and latitude. (Note that location is included as a random-effect factor as well, to capture location-based effects not present in the non-linear interaction of longitude and latitude.) A similar approach was taken by Wieling et al. [27] to model the effect of geography on Dutch dialect distances (compared to standard Dutch).

Figure 5.3 shows the resulting surface for the complete area under study using a contour plot (note that the effects of social and lexical variables are also taken into account in the model from which this surface is extracted; see Sect. 4). The (red) contour lines represent distance isoglosses connecting areas which have a similar pronunciation distance from standard Catalan. Wherever the contour lines are not regular circles, the treatment of geography is more sophisticated than in models which examined linguistic variation as a function of geographic distance alone ([40], inter alia). A green color indicates the use of pronunciations closest to the standard language, while yellow, orange, pink and light gray indicate increasingly greater pronunciation distances (on average, considering all words) from standard Catalan, respectively. The measurement points are identified by a single character corresponding to the region (A: Aragon, C: Catalonia, D: Andorra). We can clearly identify the separation between the dialects spoken in the east of Catalonia compared to the Aragonese varieties in the west. The local cohesion in Fig. 5.3 is sensible, as nearby communities tend to speak dialectal varieties which are relatively similar.

The complexity of the surface shown in Fig. 5.3 is reflected by the estimated degrees of freedom of the spline, in this case 12. The thin plate regression spline was highly significant as the 12.0 estimated degrees of freedom invested in it were supported by an F-value of 17 ($p < 0.0001$). This indicates that the non-linear surface is clearly warranted.

3.3.1 Social and Lexical Variables

In addition to the random-effect factors for word, speaker and location, and the smooth combining longitude and latitude representing geography, we considered several other predictors. Based on our initial analyses which showed that the pronunciations of articles, clitic pronouns and demonstrative adjectives (i.e. words

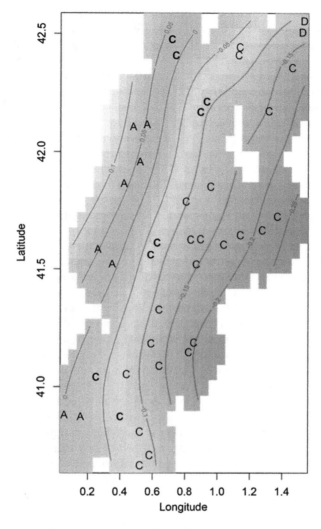

Fig. 5.3 Contour plot for the regression surface of pronunciation distance as a function of longitude and latitude obtained with a generalized additive model using a thin plate regression spline. The (red) contour lines represent (log-transformed Levenshtein) distance isoglosses, a green color (lower values, negative in the east) indicate smaller distances from the standard language, while a yellow, orange, pink and light gray color (i.e. increasingly higher values) represents greater distances. The color in the online version has been replaced by greytones in print, where darker tones indicate more standard pronunciations and lighter ones less standard pronunciations. The characters indicate the region of the measurement points (A: Aragon, C: Catalonia, D: Andorra). The C characters in boldface indicate eight sites in Catalonia, later compared to the eight sites in Aragon, discussed in Sect. 4.1

such as 'this' and 'that') differed significantly more from the corresponding standard Catalan pronunciations than the other word categories, we included a factor to distinguish these two word groups (i.e. articles, clitic pronouns and demonstrative adjectives versus verbs, neuter and personal pronouns, possessive adjectives and locative adverbs). Other word-specific variables we included were the length of the word (i.e. the number of sound segments in the standard Catalan pronunciation) and the relative frequency of vowels in the standard Catalan pronunciation of each word. In addition, we included several location-specific social variables: community size, the average community age, the average community income and the relative number of tourist beds (as a proxy for the amount of tourism). The speaker-related variables we took into account were the year of birth, the gender, and the education level of the speaker. Finally, we used a factor to distinguish speakers from Catalonia and Andorra as opposed to Aragon.

Collinearity of predictors (i.e. predictors which are highly correlated with each other) is a general problem in large-scale regression studies. In our data set, communities with a larger population tend to have a higher average income and lower average age (all $|r|$'s > 0.65). Furthermore, the articles, clitic pronouns and demonstrative adjectives were much shorter than the other words, and thus the word category factor distinguishing these types from the other words is strongly related to word length ($|r| = 0.77$). While the residualization of predictors which are highly correlated has been a popular approach, Wurm and Fisicaro [41] convincingly argued that it is not a useful remedy for collinearity. Consequently, we only included the strongest predictor from each of the two groups of related predictors.

A few numerical predictors (i.e. community size and the relative number of tourist beds) were log-transformed (i.e. instead of the original value, the logarithm of that value was used) in order to reduce the potentially harmful effect of outliers. To facilitate the interpretation of the fitted parameters of our model, we scaled all numerical predictors by subtracting the mean and dividing by the standard deviation. As indicated above, we log-transformed and centered our dependent variable (i.e. the pronunciation distance per word from standard Catalan, averaged by dividing by the alignment length). Consequently, the value 0 represents the mean log-distance, negative values a smaller distance, and positive values a larger distance from the standard Catalan pronunciation. The significance of the fixed-effect factors, covariates, and smooths was extracted from the GAM model summary.

4 Results[4]

As not all words in our data set are pronounced by every speaker, the total number of cases (i.e. word-speaker combinations) in this study is 112,608.

---

[4]The paper package associated with this paper and available at the Mind Research Repository contains all data, methods and results for reproducibility. It can be found at: http://openscience.uni-leipzig.de/index.php/mr2/article/view/46.

We fitted a generalized additive mixed-effects regression model, step by step removing predictors that did not contribute significantly to the model. Predictors which correlated highly (indicated above) were not included at the same time (i.e. population average age, population average income and population size; and word length and word category), but only the strongest predictor was included for each of the two sets of predictors (if significant). With respect to the random effects, we assessed the significance of all possible random slopes and intercepts for the random-effect factors *location*, *speaker* and *word*. We only retained random intercepts and slopes when they were associated with a significant p-value (<0.05) in the model summary. We will discuss the specification of the model including all significant predictors and random effects. The model explained 73.5% of the variation in pronunciation distances from standard Catalan. This value also incorporates the variability linked to the random-effect factors. This indicates that the model is highly capable of predicting the individual distances (for specific speaker and word combinations), providing support for our approach of integrating geographical, social and lexical variables. The main contributor (62.8%) for this good fit was the variability associated with the words (i.e. the random intercepts for word). Without random-effect factors, the fixed-effect factors explained 16% of the variation. To compare the relative influence of each of these (fixed-effect) predictors, we included a measure of effect size by specifying the increase or decrease of the dependent variable when the predictor increased from its minimum to its maximum value. The effect size of the geographical smooth was calculated by subtracting the minimum from the maximum fitted value (see Fig. 5.3). Of course, the estimates of the standardized predictors may also be used as a measure of effect size, but there is no such estimate for the effect of geography, and not all numerical predictors are normally distributed. On the basis of our measure of effect size, we clearly observe that geography and the word-related predictors have the greatest influence on the pronunciation distance from standard Catalan.

The coefficients and the associated statistics of the fixed-effect factors and covariates included in the final model are shown in Table 5.1. The random-effect factors included are shown in Table 5.2. The fact that a random intercept for location was necessary indicates that there is variability associated with the locations which is not captured by the geographical smooth. As an example of the random-effect structure, Fig. 5.4 shows the by-word random intercepts. In general, the words *cantaríeu*, *jo* and *nosaltres* are more likely to be similar to the standard Catalan pronunciations than *sentiríeu*, *canta* and *el (faran)*.

4.1 Demographic Predictors

None of the location-based predictors (i.e. the relative number of tourist beds, community size, average community income and average community age was significant as a main effect in our general model (see Table 5.1). All location-based predictors, however, showed significant word-related variation (see Table 5.2).

Table 5.1 Fixed-effect factors and covariates of the final model

	Estimate	Std. error	p-value	Effect size
Intercept	−0.033	0.018	0.061	
Vowel ratio per word	0.109	0.014	<0.001	0.674
Word category is A/D/C	0.101	0.034	0.003	0.101
Speaker year of birth (Aragon)	0.005	0.004	0.282	0.014
Speaker year of birth (Catalonia and Andorra)	−0.012	0.005	0.028	−0.034
s(longitude, latitude) [12.0 edf]			<0.001	0.310

Negative estimates indicate more standard-like pronunciations (for increasing values of the predictors), and positive estimates less standard ones. Effect size indicates the increase or decrease of the dependent variable when the predictor value increases from its minimum to its maximum value (i.e. the complete range). The geographical smooth (Fig. 5.3; 12 estimated degrees of freedom) is represented by the final row. Its effect size equals the minimum value subtracted from the maximum value of the fitted smooth

Table 5.2 Significant random-effect parameters of the final model

Factors	Random effects	Std. dev.	p-value
Word	Intercept	0.258	<0.0001
	Relative no. of tourist beds	0.025	<0.0001
	Average community age	0.031	<0.0001
	Community size (log)	0.020	<0.0001
	Average community income	0.032	<0.0001
	Speaker education level	0.009	<0.0001
	Speaker year of birth (Cat. + And.)	0.029	<0.0001
	Speaker year of birth (Aragon)	0.019	<0.0001
Speaker	Intercept	0.025	0.0004
	Vowel ratio per word	0.009	<0.0001
	Word category is A/D/C	0.018	<0.0001
	Word length	0.013	<0.0001
Location	Intercept	0.026	<0.0001
	Speaker year of birth (Cat. + And.)	0.021	<0.0001
	Vowel ratio per word	0.015	<0.0001
	Word category is A/D/C	0.071	<0.0001
	Word length	0.037	<0.0001

For example, while there is no main effect of average community income, the pronunciation of some words will be closer to the standard in richer communities, while for some other words this pattern will be reversed.

The non-linear interaction of longitude and latitude (see Fig. 5.3) shows that the Aragonese varieties have a higher distance from standard Catalan than the other varieties. In fact, if the non-linear interaction is replaced by a contrast between the Aragonese varieties versus the other varieties (also including location as a random-effect factor), the contrast is highly significant, $p < 0.0001$, and indicates that the Aragonese speakers have a larger pronunciation distance from standard Catalan than

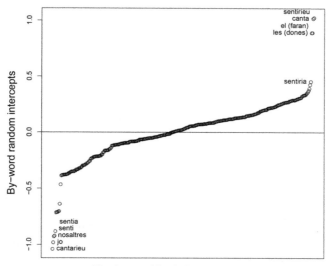

Fig. 5.4 By-word random intercepts. The words are sorted by the value of their intercept. Negative values (bottom-left) are associated with words which are generally (across all varieties) more similar to the standard, while positive values (top-right) are associated with words which are generally more different from the standard language. The dashed line shows the population intercept (see Table 5.1)

the other speakers. The same result is found when the dataset is restricted to the eight Aragonese sites and a subset of eight Catalan sites located close to the border (indicated by boldface C's in Fig. 5.3).

With respect to the speaker-related predictors, only year of birth for Catalonia and Andorra was a significant predictor, indicating that younger speakers in those two regions use pronunciations which are more similar to standard Catalan than older speakers. The effect of year of birth was not significant for Aragon, and significantly different from the effect in Catalonia and Andorra ($p = 0.02$). This result confirms the existence of a clear border effect between Aragon on the one hand, and Catalonia and Andorra on the other. We interpret this difference as the effect of the Catalan language becoming official again in the 1980s in Catalonia.

We did not find an effect of gender despite this being reported in the literature frequently (see [42] for an overview). Similarly, Wieling et al. [27] also did not find a gender effect with respect to the pronunciation distance from the standard language (Dutch) in their study. We also did not find gender differences when investigating individual linguistic variables (see Sect. 4.3, below).

We did not find support for the inclusion of education level as a fixed-effect predictor in our model. The education measure alone (without any other social status measures) might have too little power to discover social class effects ([43], Chap. 5; but see [44] for a new analysis of Labov's data suggesting that education does have sufficient power). Furthermore, when investigating individual linguistic variables (see Sect. 4.3), education only appeared once as a significant predictor.

4.2 Predictors Specific to Lexical Identity

Two variables specific to lexical identity we tested appeared to be significant predictors of the pronunciation distance from standard Catalan. It is not surprising that the binary predictor distinguishing articles, clitic pronouns and demonstratives from the other word types was highly significant, since we grouped these word categories on the basis of their higher distance from the standard language (according to our initial analyses). Articles and clitic pronouns are relatively short (in many cases only having a length of one or two sounds), and when they are different from the standard, their relative distance will be very high. While the demonstratives are not as short, they tend to be either completely identical to the standard pronunciation, or almost completely different from the standard pronunciation, which might explain their larger distances. As word length correlated highly ($|r| = 0.77$) with the binary group distinction, we only included the better predictor of the two. Given that word length was not significant, we included the binary group distinction between articles, clitic pronouns and demonstratives versus the other word types.

Finally, the number of vowels compared to the total number of sounds in the reference pronunciation was a highly significant predictor. This is not surprising (and similar to the result reported by Wieling et al. [27] for Dutch) as vowels are much more variable than consonants (e.g., [45]). Similarly to word length, including this predictor allows us to more reliably assess the effect of the more interesting predictors.

With respect to the random effects, all lexical variables showed significant variation in their strength for individual speakers and locations. This reflects that, for example, some speakers will pronounce words with a large number of vowels closer to the standard Catalan pronunciation than others.

4.3 Comparison to Individual Linguistic Variables

This paper proceeds from an aggregate, dialectometric perspective and applies a novel statistical technique, generalized additive mixed-effects regression modeling to a large collection of Catalan dialect variation data with the goal of understanding the (quite effective) standardization policies now in place in Catalonia and Andorra. The advantage of the aggregate perspective is its bird's eye view of language variation, which, in this case has meant a view encompassing over 100,000 pronunciations, 357 words (though note the lack of nouns, and the limited number of distinct verbs) as pronounced by eight speakers in each of the 40 different northwestern Catalan varieties. The aggregate perspective clearly runs the risk of losing sight of important details of language variation, but we have shown that mixed-effects regression modeling, in which words are individually modeled, can effectively detect very different levels of influence among individual words, thus protecting us against the risk of missing details, at least to some extent.

Standard sociolinguistic practice is rather different. With the goal of identifying individual phonemic changes in progress, and in particular, their social motivation, sociolinguists ignore aggregate tendencies in favor of detailed studies on the influence of social and structural factors on linguistic variation [46]. This low-level focus has certainly proven effective in understanding individual sound changes and in isolating the social dynamics that may underlie them, but it clearly runs the risk of selectively focusing on non-representative material and myopically losing sight of global tendencies.

With respect to the present study on the effects of a policy of language standardization, we might expect there to be global effects, and, in fact, this is just what we have shown. Age was shown to be significant, where the young, who have mandatorily been exposed to standard Catalan in school (and via public media), speak varieties of Catalan that are more standard like. Might we have reached similar conclusions by examining individual linguistic variables? After all, individual phoneme effects will also be reflected directly in pronunciation distances.

To answer this question, we have examined three different linguistic variables reported in the literature, to see if the effect observed at the aggregate level could also be found when focusing on a lower level. In each case we examine examples of the variables in our own data, taking care that only examples in the relevant phonetic contexts are used. Naturally we study each of them on the basis of the pronunciations of the eight speakers per site at the 40 sites described above.

The first linguistic variable (V1) we investigated was the replacement of [ʎ] (standard) by [j] (non-standard). This change has been reported by Recasens [47] and is caused by the influence of the Spanish language, from which [ʎ] has almost completely disappeared. The following 10 words present in our data set were used to examine this phenomenon: *aquell, aquella, aquells, aquelles, ell, ella, ells, elles, allò,* and *allí.*

The second linguistic variable (V2) is the variation in the final morphemes for the present subjunctive. The standard uses [i] as its subjunctive theme vowel, while other vowels indicate a non-standard pronunciation. This difference is described by Massanell [48]. We examined this variable by focusing on the following 20 items: *canti* (1[-PLU]), *cantis, canti* (3[-PLU]), *cantin, perdi* (1[-PLU), *perdis, perdi* (3[-PLU]), *perdin, begui* (1[-PLU]), *beguis, begui* (3[-PLU]), *beguin, senti* (1[-PLU]), *sentis, senti* (3[-PLU]), *sentin, serveixi* (1[-PLU]), *serveixis, serveixi* (3[-PLU]), and *serveixin.*

The final linguistic variable (V3) is the use of [β] as opposed to another consonant (mainly [w]) within the feminine possessive adjectives. The progressive substitution of [w] for the standard [β] in the Tremp area is discussed by Romero [49]. To investigate this pattern, we investigated the following six items: *meva, meves, teva, teves, seva,* and *seves.*

Table 5.3 Significance of social predictors (rows) for each of the three models corresponding each to a single linguistic variable (columns)

	V1: [ʎ] vs. [j]	V2: [i] vs. other vowel	V3: [β] vs. other consonant
Speaker is male	1.1 ($p = 0.08$)	n.s.	n.s.
Speaker education level	n.s.	−0.4*	−0.4 ($p = 0.1$)
Speaker year of birth (Catalonia and Andorra)	3.1**	−1.0**	−1.4**
Speaker year of birth (Aragon)	6.4**	n.s.	n.s.
Geography	[9.4 edf]**	[20.5 edf]**	[3.8 edf]**

Only if an estimate was significantly different from zero (or close to significance) is its estimate printed. A positive estimate indicates a greater likelihood of having a non-standard variant for increasing values of the predictor, while a negative estimate indicates the opposite. In all cases, geography shows a significant non-linear pattern (visualized in Fig. 5.5) as the edf values are greater than 1. Note that the estimates for the year of birth do not differ significantly for the two regions. Significance: *$p < 0.05$; **$p < 0.001$

Table 5.3 shows the significance of the social variables (gender, education level and age—the latter separated for the two areas) in addition to the influence of geography (visualized in Fig. 5.5). The estimates were obtained by creating three separate generalized additive mixed-effects logistic regression models (one for each linguistic variable). This approach is similar to the approach outlined in Sect. 3, except that we now use logistic regression, since in each of the three models, the dependent variable has only two values: 1 (the variant of a speaker differs from the standard language) and 0 (the variant of a speaker is equal to the standard language). In logistic regression the estimates need to be interpreted with respect to the logit scale (i.e. the log of the odds of observing a non-standard as opposed to a standard Catalan form). A positive estimate therefore indicates that an increase in the predictor results in a higher likelihood of using a non-standard variant, while a negative estimate indicates the opposite (thus the signs of the estimates can be compared to those in Table 5.1). This logistic regression approach corresponds with standard sociolinguistic practice [43].

The geographical pattern (visualized in Fig. 5.5) varies for each variable, but in general shows that the Aragonese varieties (in the west) are more likely to have a non-standard variant than the varieties in Catalonia and Andorra. Again, excluding the geographical smooth and replacing it by a binary predictor distinguishing Aragon from the other regions reveals that the Aragonese speakers are significantly more likely to use a non-standard form than the speakers from Catalonia or Andorra. The same holds when focusing on the eight Aragonese sites compared to the eight sites in Catalonia close to the border with Aragon.

With respect to the social variables, both V2 and V3 show a pattern consistent with the result presented in Table 5.1 (i.e. younger speakers are more likely to conform to the standard in Catalonia and Andorra, but not in Aragon). V1 shows that younger speakers in Catalonia and Andorra are more likely to differ from the

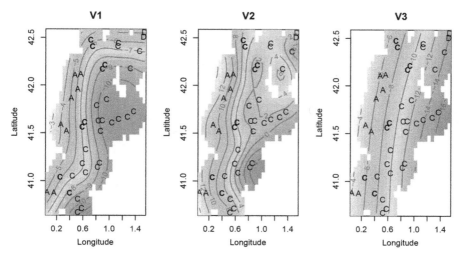

Fig. 5.5 Contour plot for the regression surfaces for each of three linguistic variables as a function of longitude and latitude obtained with a generalized additive model using a thin plate regression spline. The (red) contour lines represent isoglosses reflecting the probability (in terms of logits) of using a non-standard Catalan form, a green color (lower values in the east) indicates a smaller likelihood of using a non-standard variant, while a yellow, orange, pink and light gray color (i.e. increasingly higher values) represent a greater likelihood of using a non-standard variant. The color in the online version has been replaced by greytones in print, where darker tones indicate a smaller likelihood of using a non-standard variant, and lighter tones a greater likelihood of using a non-standard variant. The characters indicate the region of the measurement points (A: Aragon, C: Catalonia, D: Andorra). The C characters in boldface indicate eight sites in Catalonia, later compared to the eight sites in Aragon

standard language than the older speakers (caused by the move towards Spanish, as mentioned earlier), but that this effect is even stronger in Aragon (where the influence of standard Spanish is stronger). Only V2 showed a significant influence of the education level of the speaker (with more highly educated people being more likely to use the standard variant). In summary, the aggregate result with respect to year of birth is supported by two of the three individual variables.[5]

Of course, the aggregate result is not always reflected by the behavior of individual variables, and there are two reasons for this. First, the aggregate analysis shows the general pattern when taking into account the complete set of words, and it is unlikely that all individual linguistic variables exhibit this exact same pattern.

---

[5]While the precise effect of speaker's year of birth is different for both regions (Aragon, and Catalonia and Andorra) across all three variables, the difference in the effect of this predictor on Aragon as opposed to Catalonia and Andorra was never significant (all p's > 0.07) due to the small number of locations in Aragon (i.e. eight) and the limited number of words. Therefore, strictly speaking, none of the variables completely adheres to the aggregate pattern (where this difference was significant).

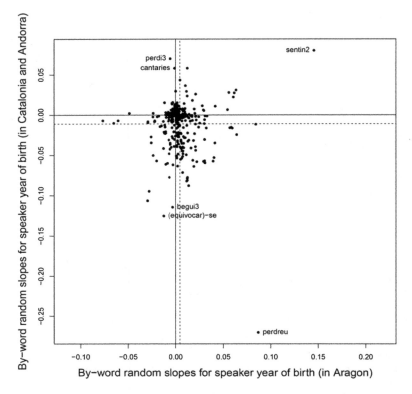

Fig. 5.6 By-word random slopes for the speaker's year of birth in Aragon (*x*-axis) and Catalonia and Andorra (*y*-axis). The dashed lines indicate the model estimates (see Table 5.1)

The second reason is that the aggregate analysis involves pronunciation distances, which also include pronunciation differences that are outside of the focus of the specifically selected linguistic variables.

By way of illustration that individual words do not all have to adhere to the aggregate pattern, Fig. 5.6 shows the by-word random slopes for the speaker's year of birth for Aragon (*x*-axis) and Catalonia and Andorra (*y*-axis). Consequently, words (i.e. dots) to the right of the *y*-axis (the vertical dashed line indicates the non-significant positive effect of speaker's year of birth for Aragon; see Table 5.1) and below the *x*-axis (the horizontal dashed line indicates the negative effect of speaker's year of birth for Catalonia and Andorra; see Table 5.1) roughly adhere to the general pattern. For words in that area, younger speakers (i.e. having a higher year of birth) in Catalonia and Andorra have a pronunciation closer to standard Catalan than older speakers, while the effect is opposite (but non-significant) in Aragon. Whereas many words follow the aggregate pattern, some words even show opposite patterns, such as *perdi3*, 'waste' (3[-PLU]). These words differ *more* from the standard for younger speakers in Catalonia and Andorra as opposed to older speakers, and differ *less* from the standard for younger people as opposed to older people in Aragon. Consequently, a linguistic variable consisting of such words would show a completely different pattern (such as V1, illustrated earlier). The aggregate approach, however, is necessary to draw more general conclusions.

5 Discussion and Conclusions

In this study we have used a generalized additive mixed-effects regression model to provide support for the existence of a border effect between Aragon (where the Catalan language does not have an official status) and Catalonia and Andorra (where Catalan is an official language). Our analysis clearly indicated a greater distance from standard Catalan for speakers in Aragon as opposed to those in Catalonia and Andorra. Furthermore, our analysis identified a significant effect of speaker age (with younger speakers having pronunciations closer to standard Catalan) for Catalonia and Andorra, but not for Aragon. This provides strong evidence for the existence of a border effect in these regions caused by different language policies and is in line with the results of Valls et al. [9]. Also, our analysis revealed the importance of several word-related factors in predicting the pronunciation distance from standard Catalan and confirms the utility of using generalized additive mixed-effects regression modeling to analyze dialect distances, with respect to traditional dialectometric analyses.

Methodologically, we have attempted on the one hand to include candidate social variables as well as geography in a single aggregate (dialectometric) analysis. We wished to include both sorts of variables in an effort to meet objections such as Woolhiser's [4] that dialectometry systematically ignores social variables. However, note that our analysis retains the aggregate perspective of dialectometry, despite the limitations caused by the data set (i.e. no nouns and only five distinct verbs). On the other hand, we have also included structural, linguistic factors in the analysis, such as the varying degree to which different words are influenced by geographic and social factors, as well as (e.g.,) the relative number of vowels in a word. Of course these linguistic techniques may seem insensitive when compared to studies in other variationist traditions (i.e. where individual sound changes are investigated), but they enable analyses to be more comprehensive, i.e. based on large amounts of data including many variables, and it has also been our point here to introduce the methodology.

With regard to the comparison to single-variable analyses, standard in sociolinguistics, we presented additional analyses at the level of three individual linguistic variables that have been discussed in the literature, and we showed that two of the three variables supported the general pattern. These analyses also illustrated that an aggregate approach is needed, as individual linguistic variables may not be representative of the global pattern.

In contrast to the (exploratory visualization-based) conclusion of Valls et al. [9] that the older speakers in urban communities use pronunciations closer to standard Catalan than the older speakers in rural communities, we did not find a significant effect of community size (nor a significant interaction between speaker age and community size). In fact when using the binary distinction Valls and colleagues based their conclusion on (i.e. distinguishing urban and rural communities in twenty different counties), the results are not at all significant ($p = 0.3$). This clearly illustrates the need for adequate statistical models, to prevent reaching statistically unsupported conclusions.

We did not find support for the importance of education level of the speaker. This might seem surprising given that one of the main reasons for the border effect is the official status of the Catalan language in both Catalonia and Andorra (and therefore its use in education), but not in Aragon. However, this education effect might be partly captured by year of birth, as there is a positive correlation between education level and the year of birth of the speaker ($r = 0.3$). Furthermore, the influence of mass media or the speaker's job might mask the potential standardizing effect of education on the speaker's pronunciation.

We also did not find support for the general influence of any of the demographic variables. This contrasts with the study of Wieling et al. [27] on Dutch dialects, who found a significant effect of community size (larger communities use pronunciations closer to the standard) and average community age (older communities use pronunciations closer to the standard language). However, the number of locations in the present study was small and might have limited our power to detect these effects—in the study of Wieling et al. [27] more than ten times as many locations were included.

It should be clear that we think that the standardization policy has led to pronunciation change. We have asked ourselves whether our reasoning commits the fallacy known as *post hoc, ergo propter hoc*—i.e. whether we might be mistaking a mere correlation between standardization policy and pronunciation change for a causal relation between the two. The temporal order is indeed as it should be, i.e. the behavioral change followed the policy change with younger people in Catalonia (where Catalan was used in schools and public media again after Franco's dictatorship) speaking a more standard-like dialect. Nonetheless, the relation might also be indirect, i.e. the policy change might have influenced attitudes which in turn influence phonetic behavior. And it is also possible that the policy change was motivated by linguistic ideology, but it would take us too far afield to explore those issues here. We admit therefore that we cannot claim to have proven that the policy change caused the pronunciation change, even if that is our interpretation.

We see three promising extensions of this study. First, replicating this study using new material (i.e. using a random set of words) would be useful to see if the results on the basis of our study (with a biased set of items) are valid in general.

Second, it would be interesting to investigate standardization towards Spanish, by comparing the dialectal pronunciations to the Spanish standard language instead of the Catalan standard language. In our data set there are clear examples of the usage of a dialectal form closer to the standard Spanish pronunciation than to the standard Catalan pronunciation, and it would be rewarding to investigate which word- and speaker-related factors are related to this.

The third extension involves focusing on the individual sound correspondences between Catalan dialect pronunciations and pronunciations in standard Catalan. These sound correspondences can easily be extracted from the alignments generated by the Levenshtein distance algorithm. When focusing on a specific set of locations (e.g., the Aragonese locations), it would be computationally feasible to create a generalized additive mixed-effects regression model to investigate which factors determine when a sound in a certain dialectal pronunciation is different from the corresponding sound in the standard Catalan pronunciation.

Acknowledgements We thank the two anonymous reviewers for their extensive comments which have helped to improve this manuscript. This research was partly funded by the project *Descripción e interpretación de la variación dialectal: aspectos fonológicos y morfológicos del catalán* (FFI2010-22181-C03-02), financed by MICINN and FEDER.

References

1. Woolard K, Gahng T-J (2008) Changing language policies and attitudes in autonomous Catalonia. Lang Soc 19:311–330
2. Huguet A, Vila I, Llurda E (2000) Minority language education in unbalanced bilingual situations: A case for the linguistic interdependence hypothesis. J Psycholinguist Res 3:313–333
3. Fabra P (1918) Gramàtica catalana. Institut d'Estudis Catalans, Barcelona
4. Woolhiser C (2005) Political borders and dialect divergence/convergence in Europe. In: Auer P, Hinskens F, Kerswill P (eds) Dialect change. Convergence and divergence in European languages. Cambridge University Press, New York, pp 236–262
5. Goebl H (2000) Langues standards et dialectes locaux dans la France du Sud-Est et l'Italie septentrionale sous le coup de l'effet-frontière: une approche dialectométrique. Int J Sociol Lang 145:181–215
6. Pradilla M-À (2008) Sociolingüística de la variació i llengua catalana. Institut d'Estudis Catalans, Barcelona
7. Pradilla M-À (2008) La tribu valenciana. Reflexions sobre la desestructuració de la comunitat lingüística. Onada, Benicarló
8. Bibiloni G (2002) Un estàndard nacional o tres estàndards regionals? In: Joan B (ed) Perspectives sociolingüístiques a les Illes Balears. Res Publica, Eivissa
9. Valls E, Wieling M, Nerbonne J (2013) Linguistic advergence and divergence in north-western Catalan: A dialectometric investigation of dialect leveling and border effects. LLC J Digit Scholarsh Humanit 28(1):119–146
10. Bailey G, Wikle T, Tillery J, Sand L (1991) The apparent time construct. Lang Var Chang 3:241–264
11. Wieling M (2012) A quantitative approach to social and geographical dialect variation. PhD thesis, Rijksuniversiteit Groningen
12. Nerbonne J (2009) Data-driven dialectology. Lang Ling Compass 3(1):175–198
13. Wieling M, Nerbonne J (2011) Bipartite spectral graph partitioning for clustering dialect varieties and detecting their linguistic features. Comput Speech Lang 25(3):700–715
14. Wieling M, Nerbonne J (2015) Advances in dialectometry. Ann Rev Linguist 1:243–264
15. Institut d'Estudis Catalans (1999) Proposta per a un estàndard oral de la llengua catalana I. Fonètica. Institut d'Estudis Catalans, Barcelona
16. Institut d'Estudis Catalans (1999) Proposta per a un estàndard oral de la llengua catalana II. Morfologia. Institut d'Estudis Catalans, Barcelona

17. Institut d'Estadística de Catalunya (2008, 2010) Territori. http://www.idescat.cat. Accessed 28 Feb 2011
18. Instituto Aragonés de Estadística (2007, 2009, 2010) Población y Territorio. http://www.aragon.es. Accessed 28 Feb 2011
19. Departament d'Estadística del Govern d'Andorra (2010) Societat i població: http://www.estadistica.ad. Accessed 28 Feb 2011
20. Cambra de Comerç – Indústria i Serveis d'Andorra (2008) Informe econòmic
21. Levenshtein V (1965) Binary codes capable of correcting deletions, insertions and reversals (in Russian). Dokl Akad Nauk SSSR 163:845–848
22. Heeringa W (2004) Measuring dialect pronunciation distances using Levenshtein distance. PhD thesis, Rijksuniversiteit Groningen
23. Heeringa W, Kleiweg P, Gooskens C, Nerbonne J (2006) Evaluation of String Distance Algorithms for Dialectology. In: Nerbonne J, Hinrichs E (eds) Linguistic distances workshop at the joint conference of International Committee on Computational Linguistics and the Association for Computational Linguistics, Sydney, pp 51–62
24. Wieling M, Prokić J, Nerbonne J (2009) Evaluating the pairwise string alignment of pronunciations. In: Borin L, Lendvai P (eds) Language technology and resources for cultural heritage, social sciences, humanities, and education, Workshop at the 12th Meeting of the European Chapter of the Association for Computational Linguistics. Athens, 30 March 2009, pp 26–34
25. Church K, Hanks P (1990) Word association norms, mutual information, and lexicography. Comput Linguist 16(1):22–29
26. Wieling M, Margaretha E, Nerbonne J (2012) Inducing a measure of phonetic similarity from pronunciation variation. J Phon 40(2):307–314
27. Wieling M, Nerbonne J, Baayen RH (2011) Quantitative social dialectology: explaining linguistic variation socially and geographically. PLoS One 6(9):e23613
28. Wieling M, Bloem J, Mignella K, Timmermeister M, Nerbonne J (2014) Measuring foreign accent strength in English: Validating Levenshtein distance as a measure. Lang Dyn Change 4(2):253–269
29. Tagliamonte S, Baayen RH (2012) Models, forests and trees of York English: was/were variation as a case study for statistical practice. Lang Var Chang 24(2):135–178
30. Baayen RH, Davidson DJ, Bates DM (2008) Mixed-effects modeling with crossed random effects for subjects and items. J Mem Lang 59(4):390–412
31. Barr DJ, Levy R, Scheepers C, Tily HJ (2013) Random effects structure for confirmatory hypothesis testing: Keep it maximal. J Mem Lang 68(3):255–278
32. Bates D, Kliegl R, Vasishth S, Baayen RH (2015) Parsimonious mixed models. http://arxiv.org/abs/1506.04967
33. Jaeger F (2008) Categorical data analysis: away from ANOVAs (transformation or not) and towards logit mixed models. J Mem Lang 59(4):434–446
34. Baayen RH (2008) Analyzing linguistic data. A practical introduction to statistics using R. Cambridge University Press, Cambridge
35. Hastie TJ, Tibshirani RJ (1990) Generalized additive models, vol 43. CRC Press, Boca Raton
36. Wood S (2006) Generalized additive models: An introduction with R. Chapman & Hall/CRC, Boca Raton
37. Wood S (2003) Thin plate regression splines. J R Stat Soc Ser B (Stat Methodol) 65(1):95–114
38. Wood S (2011) Fast stable restricted maximum likelihood and marginal likelihood estimation of semiparametric generalized linear models. J R Stat Soc Ser B 73(1):3–36
39. Nerbonne J (2010) Measuring the diffusion of linguistic change. Philos Trans R Soc B Biol Sci 365:3821–3828
40. Nerbonne J, Heeringa W (2007) Geographic distributions of linguistic variation reflect dynamics of differentiation. In: Featherston S, Sternefeld W (eds) Roots: Linguistics in search of its evidential base. Mouton De Gruyter, Berlin, pp 267–297
41. Wurm LH, Fisicaro SA (2014) What residualizing predictors in regression analysis does (and what it does not do). J Mem Lang 72:37–48

42. Cheshire J (2002) Sex and gender in variationist research. In: Chambers JK, Trudgill P, Schilling-Estes N (eds) The handbook of language variation and change. Blackwell, Oxford, pp 423–443
43. Labov W (2001) Principles of linguistic change, volume 2. Social factors. Blackwell, Oxford
44. Gorman K (2010) The consequences of multicollinearity among socioeconomic predictors of negative concord in Philadelphia. In: Lerner M (ed) University of Pennsylvania Working Papers in Linguistics, vol 16(2), pp 66–75
45. Keating P, Lindblom B, Lubker J, Kreiman J (1994) Variability in jaw height for segments in English and Swedish VCVs. J Phon 22:407–422
46. Chambers J (2009) Sociolinguistic theory: Linguistic variation and its social significance, 3rd edn. Wiley-Blackwell, Oxford
47. Recasens D (1996) Fonètica descriptiva del català (assaig de caracterització de la pronúncia del vocalisme i consonantisme del català al segle XX). Institut d'Estudis Catalans, Barcelona
48. Massanell M (2001) Morfologia flexiva actual de la Seu d'Urgell i Coll de Nargó: Estadis en el procés d'orientalització del català nord-occidental. Z Katalan 14:128–150
49. Romero S (2001) Canvi lingüístic en morfologia nominal a la Conca de Tremp. Universitat de Barcelona. PhD thesis. Available at http://hdl.handle.net/10803/2082

Chapter 6
Evaluating Logistic Mixed-Effects Models of Corpus-Linguistic Data in Light of Lexical Diffusion

Danielle Barth and Vsevolod Kapatsinski

Abstract We explore methods for evaluating logistic mixed-effects models of both corpus and experimental data types through simulations. We suggest that the fit of the model should be evaluated by examining the variance explained by the fixed effects alone, rather than both fixed and random effects put together. Nonetheless, for corpus data, in which frequent items contribute more observations, coefficient estimates for fixed effects should be derived from a model that includes the random effects. Including random effects in the model with such datasets allows for better estimates of the fixed-effects predictor coefficients. Not having random effects in the model can cause fixed-effects coefficients to be overly influenced by frequent items, which are often exceptional in linguistic data due to lexical diffusion of ongoing changes.

1 Mixed-Effects Models in Corpus Linguistics

Linguistics is fundamentally concerned with explaining why people say what they say when they say it. Given multiple ways of conveying the same message [1, 2], what makes speakers choose one way over another in a particular situation? The fact that there are multiple ways of conveying roughly the same intended meaning is easy to see in the case of a multilingual speaker: she might be able to say "I am a linguist" in multiple languages and would choose the language appropriate to the situation. However, it is likewise true for monolinguals. For instance, part of what causes speakers to choose *the wheel of the car* over *the car's wheel* has been shown to be the fact that the latter option results in two stressed syllables being placed next to each other [3]. Of course, this avoidance of stress clash is not the only predictor influencing the choice between the two genitive constructions above. Typically, many factors, semantic, syntactic, phonological, social, etc. impact

---

D. Barth · V. Kapatsinski (✉)
University of Oregon, Eugene, OR, USA
e-mail: vkapatsi@uoregon.edu

© Springer International Publishing AG 2018
D. Speelman et al. (eds.), *Mixed-Effects Regression Models in Linguistics*,
Quantitative Methods in the Humanities and Social Sciences,
https://doi.org/10.1007/978-3-319-69830-4_6

the choice. This multiplicity of influences has led linguists interested in predicting choice of expression in production, starting with Cedergren and Sankoff [4], to turn to multiple regression models, now the main workforce of the highly related fields of quantitative corpus linguistics and quantitative/variationist sociolinguistics (e.g., [5]), both of which deal with analyses of representative databases of natural speech, or corpora.

While traditionally these regression models had only fixed-effects predictors (e.g., [6]), mixed-effects models have now become the new standard [7–10]. One reason [9, 10], is that a valid replication of a corpus-linguistic study would not necessarily have to be a study of the same speakers, as long as the speakers come from the same speech community. Nonetheless, speakers vary around the community norm, and the effects of such variation should be taken into account, suggesting that speaker identity should be included in models of linguistic behavior as a random effect.

Linguistic items are another source of variability. As Sapir [11] pointed out, "all grammars leak": no grammar is completely predictive of linguistic behavior, and part of the divergence between the grammar and the behavior comes down to the existence of exceptional linguistic items. Even if most words containing a linguistic structure behave consistently (undergo a rule affecting that structure with some constant probability), there are usually a few exceptional words that contain the structure but do not undergo the rule. This situation is typical in all languages because speakers of all languages have abundant long-term memory, allowing them to store and retrieve frequent words and phrases like *I don't know* as wholes, rather than deriving them from their parts using the grammar (e.g., [12]). Long-term memory storage and retrieval allows these frequent phrases to become exceptional in various ways. For instance, *I don't know*, unlike other *I don't Verb* phrases, can be reduced to little more than a pattern of intonation superimposed on a nasal sound [13].

Clark [14] and Coleman [15] argued persuasively that items should be treated as a random effect in psycholinguistic experiments, since the researcher samples the items from a larger population and would like to generalize to the population rather than just the sampled items. The argument also holds for studies aimed at investigating grammar on the basis of corpus data, since grammatical generalizations are usually intended to apply across lexical items. However, corpus data differs from psycholinguistic data in two ways, which raise additional questions. First, in psycholinguistic data, every item is usually presented to every subject the same number of times. In corpus data, the items that are more frequent in the language will be observed more frequently. Second, lexical diffusion theory claims that high frequency of use leads to articulatory reduction and semantic bleaching of the frequently used item, as well as retention of grammatical patterns that are no longer productively applied to novel items [12, 16–22]. To the extent that this is true, high-frequency items are not just more likely to deviate from the sample mean than low-frequency items; they are also likely to deviate from the mean in different directions than low-frequency items. Thus in a corpus study, the sample is biased to oversample items that are likely to be exceptional. We describe these challenges in more detail in the following section. The rest of the paper is devoted to addressing the implications of these challenges for the use of mixed-effects models in corpus linguistics by means of Monte Carlo simulations.

2 The Challenges of Corpus Data Given Lexical Diffusion

As discussed above, one of the main challenges of corpus data is that the data are not nicely balanced. One source of imbalance is corpus design: unless special care is taken, more talkative (or popular) speakers will contribute more to the database than less talkative (or less popular) ones. However, even in a corpus that has been intentionally balanced by speaker (e.g. [23]), a sample from the corpus designed to investigate the use of a certain linguistic structure is still usually not balanced by speaker (e.g. [24]), as some speakers will use the structure more than others, both because of personal preference [25] and the topics being discussed.

An even more pervasive source of imbalance is the fact that any corpus contains a small number of extremely frequent words, while most words occur only once [26, 27]. Lack of balance across linguistic items is unavoidable because linguistic expressions are subject to a rich-get-richer effect. The more often a word is used, the more likely it is to be re-used in the future: frequent words come to mind more readily than infrequent words in language production (as demonstrated by Oldfield and Wingfield [28], in a picture naming task). As we would expect from a variable subject to rich-get-richer positive feedback loops [29–32], word frequencies display a power-law distribution [26, 27, 33].

Lexical diffusion theory suggests that grammatical changes spread through the lexicon in a word-by-word fashion, starting either from low-frequency or high-frequency words (see esp. [12, 20]). In particular, reductive changes motivated by simplifying or streamlining articulation are argued to start in frequently used words. These include all processes involving reducing the magnitudes of or increasing the overlap between articulatory gestures, including consonant deletion, vowel shortening and vowel centralization. For example, Bybee [17] documents that word-final *t*/*d* deletion in English is more likely in frequent words like *missed* than in rare ones like *maced*. In contrast, innovative non-reductive pronunciations and morphological processes might spread from rare words to frequent words, so that synchronically the rare words are more likely to contain the innovative variant (e.g. [20, 34]). A classic example is that the irregular past tense formation patterns in English are maintained in frequent verbs (*drink-drank*), while the rarer ones take the innovative *-ed* suffix (*wink-winked*, **wink-wank*).

However, word frequency does not account for all differences among words (e.g., [35]). Therefore, there is widespread agreement that the identity of a particular word is an additional important predictor in the study of grammar and language change. Perhaps, the least controversial example is that verbs differ in their preferences for various syntactic constructions ([36–38]; see also [39], reviewed below, for a similar case).

Fixed-effects only models (e.g. [4, 6]) have difficulty incorporating an effect of word into the model because such models have to choose between fully pooling data across words or not pooling data across words at all (as discussed in [40]). If the data are pooled across words (by not including word identity into the model as a predictor), then the predictions of the model are overly influenced by the frequent words, which contribute a large proportion of observations to the dataset. This is especially problematic in examining the effects of grammatical predictors, since frequent words are likely to be exceptional in various ways, obeying grammatical generalizations that are no longer productively applied to novel inputs, and not obeying ones that are just coming into the language (unless they are reductive in nature). Thus, fixed-effects only models excluding an effect of word are likely to overestimate the productivity of reductive patterns like *t/d* deletion and to underestimate the productivity of non-reductive innovations like the *–ed* past tense suffix.

While pooling data across words is problematic, not pooling across words at all is also problematic because there are never enough observations to estimate the effects of individual rare words. To see why, consider the fact that a significant proportion of observations in every corpus is contributed by *hapax legomena*, words that occur only once in the corpus. It is impossible to estimate the effects of individual hapax legomena, as their effects are indistinguishable from observation-level noise. Therefore *some* pooling of data across words is necessary. Baayen [26] shows that the proportion of hapax legomena in a corpus stays relatively constant as corpus size increases, suggesting that this problem will not be solved by using larger corpora or larger samples.

As Bresnan et al. [7], Gerard et al. [41], and Sonderegger [42], among others, note, mixed-effects models are intended to address this issue by *partially* pooling the data across words. We can estimate the effect of a rare word on the observed behavior based on the effects of more frequent words, which can be estimated more reliably. On the other hand, the effects of individual frequent words can be estimated without relying on data from other words.

There is some psycholinguistic motivation for partial pooling, since the problem of data sparsity that partial pooling is intended to solve is not just a problem for linguists analyzing a corpus. It is also a problem for language learners. Thus Stefanowitsch [43] notes that we can be much more confident about the intransitivity of the frequent verb *disappear* than about the intransitivity of the infrequent *vanish* (**He disappeared it.* is judged as being less grammatical than **He vanished it.*) Ambridge et al. [44, 45] show that speakers are in fact more confident about the ungrammaticality of the former, suggesting that they are more confident in transitivity of *disappear* than that of *vanish*. Frequency influences our ability to learn about the idiosyncrasies of an item [12, 17, 19, 39, 46, 47], making the between-item differences more pronounced among high-frequency words. For instance, Erker and Guy [39] show that some high-frequency verbs in Spanish favor the omission of subject pronouns while others disfavor it, whereas all low-frequency verbs are alike in this respect. Raymond and Brown [48] show that fricative reduction is exceptionally productive in some high-frequency words (ones that tend to occur in casual speech) and exceptionally unproductive in others (words that tend to occur in more formal registers). As noted by Kapatsinski [47], these results are very much in line with partial pooling: the lowest-frequency words cannot be exceptional because language learners cannot estimate individual probabilities (of co-occurring with some construction or undergoing some change) for those words and must instead rely on the lexicon-wide probabilities, which are based on data pooled across individual words.

Nonetheless, the fact is that even partial pooling assumes that low-frequency words should behave like high(er)-frequency words. However, lexical diffusion theory suggests that low-frequency words should *not* behave like high-frequency words, being more likely to undergo reductive processes and less likely to undergo non-reductive ones [12, 17, 20, 34]. Thus the items for which we have much data may not behave like the items for which the data are sparse. Given this possibility, even the simplest model of grammatical behavior should include item frequency, and not just item identity, as a predictor. This special status of frequency is why we focus on this predictor in the present paper.

3 Purposes of Model Evaluation

To the extent that a corpus is a representative sample of the recorded speakers' productions, the within-speaker predictors of a regression model of some behavior in that corpus, along with the associated coefficients, can be thought of as a (partial) description of the speakers' production grammar. Therefore evaluation and comparison of alternative regression models is fundamental to the linguistic enterprise [1, 4]. Model evaluation is concerned with two related questions:

1. how much room is there for improvement over the current model, i.e., should we look for additional predictors or have we described the grammar fully?, and
2. if two models differ in that one includes an additional predictor, which of the two models is better (and by how much)?

The answer to the first question is important for research planning. Should we try to incorporate additional influences into the model, which often involves laboriously coding the data for additional variables? Or is there not enough variance left to try accounting for? The answer to the second question is important for model selection, a crucial task if one believes in the existence of a single true model that has generated the observed behavior. If one does not, and wishes to avoid model selection by employing model averaging, the answer to the second question remains important for model weighting [49, 50].

While it is possible to employ different methods to answer the two questions, we argue that the answer to both questions should be based on how much variance is accounted for by the fixed effects predictors alone, rather than by the full model including random effects. Thus, if one employs the R^2 measures of fit for mixed-effects models recently developed by Nakagawa and Schielzeth [51], the appropriate measure to use is marginal R^2, rather than conditional R^2. If one employs resampling techniques we use here, one has no choice but to evaluate the predictiveness of a model based only on the predictiveness of its fixed effects because the model is tested on items it was not trained on.

4 The Problem

In current mixed-effects modeling of linguistic data, it is common to evaluate a model by measuring how well the full model (including random effects) fits the training data (e.g. [52], p. 281; [53–56]). This can be done using various measures of fit like Somers' Dxy, C (index of concordance), log likelihood and conditional R^2. Here, we use C but the problem would arise with any alternative measure, as long as random-effects predictors are included in the evaluated model and the model is tested on data with levels of random-effects predictors the model was fit to/trained on.

We show that the accuracy of a mixed-effects model is maintained even when the values of a real fixed-effects predictor are randomly scrambled, due to the estimated contribution of a random-effects predictor. When the fixed-effects predictors included in a model do not do a good job, a random-effects predictor can rise to the occasion and capture the same variance.

We are not the first to make this argument in linguistics. Antić [57, 58] and Yao [59] have likewise noticed this anecdotally and proposed that the goodness of a fixed-effects predictor is verified if it can capture some of the variance that would otherwise be attributed to a random effect. In other words, a more complex model can be accepted over a less complex model if the more complex model uses fixed effects to capture some of the variance that the simpler model attributes to random effects.

For example, Antić [58] found that detecting a prefix was easier, with respect to reaction time, when the word containing the prefix could be easily decomposed into morphemes. She compared two mixed-effects models, one containing measures of compositionality, her theoretically motivated fixed effects, and one that did not contain them. Adding the measures of compositionality to the model did not make the model have a better fit to the data but the measures accounted for 88% of the variance that the model without these measures attributed to the random effect of item. She concluded that compositionality does have an effect on prefix detection reaction times, a conclusion that depends on the random effects "stepping in" to capture residual variance when the fixed effects are not there to account for it.

We show that a random-effects predictor does indeed capture more variance when the values of a useful fixed-effects predictor are scrambled (randomly reordered). This can result in a model that achieves a good fit to the training data but does so by fitting the data using random effects. When a model achieves good fit to the training data using only random effects, it is essentially useless to generalize beyond the observed sample of words and/or speakers. The researcher should therefore continue searching for additional fixed-effects predictors. However, under the approach above, the best model from the set of models currently under consideration may be evaluated as highly predictive, prematurely terminating the search for additional predictors (due to incorrectly answering Question 1).

Similarly, a model containing additional fixed-effects predictors (the bigger model) may be rejected in favor of a model with fewer predictors (the smaller model) based on the two models fitting the training data equally well. We argue that this is an incorrect answer to Question 2 when the fixed-effects predictors of the bigger model capture variance that the smaller model captures using random effects.

This raises the question of whether there is any reason at all to include the random effects in the model: on the Antić/Yao approach, a fixed-effects predictor is accepted if it covaries with the dependent variable whether or not random effects can capture the same variance. We argue that this decision procedure is correct. However, we also argue that random effects predictors are nonetheless useful with unbalanced data typical of corpora for deriving predictive values of the coefficients associated with fixed-effects predictors. In other words, accounting for the idiosyncrasies of individual words allows one to build a more predictive grammar, one that shows better generalization to novel words.

Following Pitt and Myung [60], among others, we test our models on unseen data, a practice attested but uncommon in the corpus literature [61–63]. We fit the model to data containing a random subset of the levels of a random-effects predictor and test on the rest of the data. In this case, the model can only do a good job predicting the values of the dependent variable in the test data by means of fixed effects: the levels of the random effects in the test data are unfamiliar from training. On this measure, a real fixed effects predictor performs better than its randomly scrambled counterpart. The model tested on unseen data has only coefficients associated with fixed effects. However, we show that for highly unbalanced datasets typical of corpus data, the coefficient estimates are much better (in that they are more predictive) when the random effect is included in fitting the model. These results reaffirm the usefulness of mixed-effects modeling for corpus data [9, 10].

5 Simulations

5.1 Documenting the Problem

We created one thousand replications of a simple corpus study, in which there is one fixed-effects predictor and one random-effects predictor influencing the probability of reduction. The random-effects predictor was the identity of the word, while the fixed-effects predictor was word frequency. In every replication, both predictors had a real influence. The value of the dependent variable is determined by the value

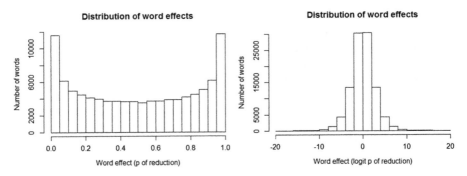

Fig. 6.1 Distribution of word effects in terms of probability of choosing the 'reduced' value of the dependent variable (bimodal) and the corresponding logits (normal) for the words in the simulations

of the fixed-effects predictor (word frequency), plus the idiosyncratic effect of the individual word (item-level noise) and random observation-level noise. For each replication, we created a new dataset by taking a new sample of frequencies from the distribution in Fig. 6.3, as well as new samples from random distributions of item-level and observation-level noise, which together define the values of the dependent variable. The script file for this simulation is available at https://www.dropbox.com/sh/7ablw99cg94vr7r/AAC-oYIMWEfZLYrKm_wrJmd9a?dl=0

The effect of word is shown in Fig. 6.1: some words are associated with one value of the dependent variable while others favor the other value. The distribution of probabilities is bimodal, as it should be, given the frequent observation that speakers may be uncertain about the grammatical behavior of unknown words and yet be certain about the behavior of the words they know (e.g., [19, 46, 64]). Nonetheless, the distribution of the corresponding logits (log-odds) is quasi-normal: if we compare the distribution for every replication to a normal distribution with the same number of items, mean and standard deviation using the Kolmogorov-Smirnov test, it comes out significant ($p < 0.05$) only 2.9% of the time in the sample of 1000 replications. Two normal distributions with the same mean, standard deviation and number of observations come out as significantly different 3.2% of the time in the sample. Thus, the distribution of item effects is quasi-normal in logit space, making logistic regression appropriate for analyzing the data.

For every replication, we ran two mixed-effects models: one model included the real fixed-effects predictor while the other included its randomly scrambled version. Scrambling was done between items. In other words, in both models with original frequency values and models with scrambled frequency values, frequency was constant across observations of a single linguistic item. Thus the hierarchical structure of the dataset was preserved. Each replication used a different random scrambling of the values of the real predictor.

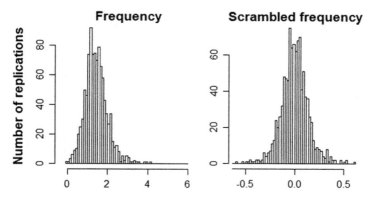

Fig. 6.2 The distribution of the effect of frequency (left) vs. scrambled frequency (right)

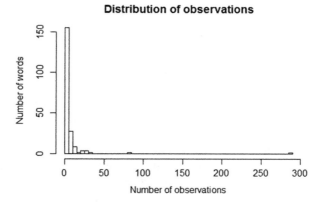

Fig. 6.3 The distribution of number of observations across words for a corpus-like sample, in which number of observations for a word is proportional to its frequency

Models were fit using the lme4 package in R (version 0.999999-0, [65]). By comparing the two models to each other, we can then attempt to determine the predictive power of the fixed-effects predictor. Figure 6.2 shows that the manipulation worked: the fixed effect of frequency was greater than the fixed effect of scrambled frequency: the coefficients for frequency center around 1.5 while those for scrambled frequency center on zero.

For every one of the 1000 replications, two versions of the dataset were created. In one version, the number of observations for each word was directly proportional to its frequency: frequent words contributed more observations. Frequency (and hence number of observations) was distributed according to the Pareto distribution [33], illustrated in Fig. 6.3. This version is more likely to resemble the samples obtained in a corpus study, where more observations are found of frequent items (e.g., [7]). In the other version, each word contributed the same number of

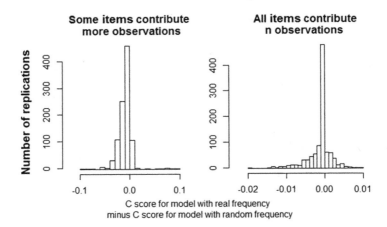

Fig. 6.4 Histograms of differences in C scores between a model that has real frequency values and a model that contains scrambled frequency values as a predictor. The C scores for the model containing the real predictor are no greater than the C scores of the control model that contains scrambled frequency values as a predictor

observations, which was equal to the mean number of observations contributed by a word in the other condition. (Thus, mean number of observations per word was equated in the two sampling conditions). The balanced sampling scheme is usual in experimental studies.

The first way we compared the models was using the Concordance index on the entire dataset to which the models were fit, leaving the random-effects predictors in the model (as suggested in [52]). The distributions of concordance indices for balanced sampling and skewed sampling are shown in Fig. 6.4. As Fig. 6.4 shows, the distributions are virtually identical: scrambling the frequency predictor did not decrease the index of concordance of the model despite reducing the coefficient associated with the fixed-effects predictor to zero. Because the coefficient associated with the scrambled fixed-effects predictor is zero, the scrambled fixed-effects predictor is useless for predicting the dependent variable. Nonetheless, the model containing only the scrambled fixed effect plus random effects appears to be as good as a model containing the real fixed effect and the same random effects.

This result confirms the hypotheses of Antić [57, 58] and Yao [59] that a real predictor can nonetheless fail to contribute to how well a mixed-effects model fits the data. The random-effects predictor steps in to capture the variance that the fixed-effects predictor is no longer capturing once it is scrambled. The result holds for both corpus data, where number of observations is correlated with values of the fixed-effects predictor in question, frequency, and experimental data where the two are uncorrelated and the number of observations per cell in the design is controlled. We believe that this result conclusively argues against using fit of the complete model (including random effects) to evaluate mixed-effects models.

5.2 The Solution: Using Mixed-Effects Models to Derive Coefficients of the Evaluated Fixed-Effects Models When the Sample is Unbalanced

In the remaining simulations we turn to comparing the models on predicting the data they have not been fit to by means of leave-one-out cross-validation (LOOCV). We believe this is a good way to determine how predictive the different models are (e.g. [60]) although other ways of testing predictiveness, including AIC scores [50] and marginal R^2 [51], would achieve the same aim.

On every replication, we again derived a new dataset in the same way as before. We then cycled through all the items, training the models on all items but one and testing it on the remaining, withheld, item. As before, we compared models containing real frequency values to models containing scrambled ones. Since the test item was withheld from training, the models' random-effects predictor 'item' does not have a level corresponding to the test item. Thus, for every mixed-effects model, we extracted the intercept and the coefficient associated with the fixed effect of frequency and generated predicted values for the unseen item using those coefficients and the observed frequency value of the unseen item.

The model we are testing now on unseen data is a fixed-effects-only model: the only predictor is frequency, which is a fixed effect. This is the crucial difference from the problematic method described in the previous section. The difference is analogous to using marginal R^2 rather than conditional R^2 in Nakagawa and Schielzeth [51].

While the evaluated model has only the fixed-effects predictor, the coefficient for that predictor was estimated using the mixed-effects model. We reasoned that the resulting estimate may be more accurate than one that would be obtained by fitting a fixed-effects-only model to the training data because it would partial out variability due to individual items. To test this idea, we also generated a fixed-effects-only (near-)equivalent to every mixed-effects model using the glm() function in R. The fixed-effects-only model had no random-effects predictors.

Figure 6.5 shows how well the resulting models fit the test data (measured by a difference in prediction accuracy, i.e. accuracy on the test items withheld from training) compared to models that contained scrambled frequency values. Each point represents a different replication. If the real predictor allows the model to achieve better prediction accuracy, we should expect the difference in accuracies to be positive. The data suggest that having the random effect in the model that is fit to the training data makes the coefficient associated with the fixed effect more accurate when the number of observations varies, as in corpus studies.

The GLM and GLMM perform very similarly for the balanced sample: the right (unscrambled) predictor is preferred by both models; model prediction accuracies across replications correlate at $r = 0.95$, and both models achieve better prediction accuracy with the unscrambled predictor, and therefore prefer the right predictor, approximately 90–91% of the time ($\chi^2(1) = 0.28$, $p = 0.6$). However, the mixed-effects GLMM is much more likely to achieve better prediction accuracy with the

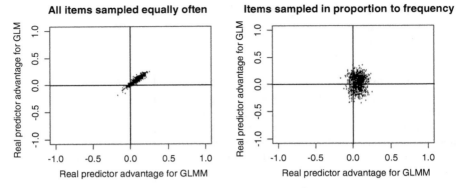

Fig. 6.5 Mixed-effects (GLMM) vs. fixed-effect-only (GLM) models and sampling. Negative values indicate that the model achieves higher accuracy with the scrambled frequency values while positive ones indicate that it achieves higher accuracy with real frequency values. With a balanced (experiment-like) dataset to be trained on, GLM and GLMM perform similarly; with corpus-like sampling, the GLMM is superior in distinguishing the real predictor from the scrambled predictor

real frequency values when number of observations varies along with frequency: 87% for GLMM vs.62% for GLM ($\chi^2(1) = 154.3$, $p < 0.00001$). We should note that corpus-like sampling reduced the likelihood of achieving better prediction accuracy with real frequency compared to scrambled frequency even for the mixed-effects model (91% vs. 87%; $\chi^2(1) = 7.59$, $p = 0.006$) but the mixed-effects model shows a much better ability to cope with corpus-like sampling. These data suggest that fixed-effects coefficient estimates are best estimated using mixed-effects models for corpus data.

What is it then about the corpus-like data that causes fixed-effect coefficients estimated on the basis of mixed-effects models to be so superior to those estimated by the GLM? In particular, is it crucial that number of observations is correlated with the values of the fixed-effects predictor in question? Is it crucial that predicted values are biased in favor of the value of the dependent variable associated with values of the fixed-effects predictor for which we have more observations? Is it crucial that distinct observations of an item always have the same value on the fixed-effects predictor? Or is the fact that different items contribute different numbers of observations sufficient to make the coefficient estimates based on the mixed-effects model superior? We argue that unbalanced sampling across levels of the random-effects predictor is sufficient, hence the superiority of coefficient estimates based on mixed-effects models should be true for all kinds of predictors in corpus studies.

To address this question, we switched the actual fixed-effects predictor to be uncorrelated with frequency and item by randomly selecting the level of the predictor (either "1" or "0") for every observation. As before, the value of the dependent variable for an observation was generated from the value of the predictor, plus item-level and observation-level noise. The dependent variable was then a function of both item identity and the value of the predictor but the value of the predictor varied randomly within and across items. In this new simulation, the fixed-

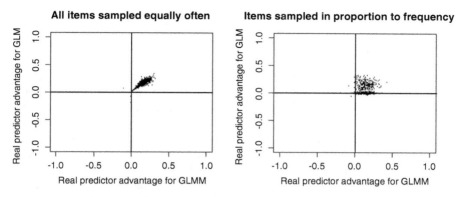

Fig. 6.6 Mixed-effects (GLMM) vs. fixed-effect-only (GLM) models and sampling. For balanced samples, both models work very similarly and always prefer the real predictor; for corpus-like samples, the fixed-effects-only model (GLM) is much more likely to fail to prefer the real predictor whereas the mixed-effects model (GLMM) continues to almost always prefer the actual predictor

effects predictor differed from the fixed-effects predictor in previous simulations in all potentially relevant respects: (1) it was binary rather than continuous, (2) the value of the predictor was not correlated with number of observations of that value, (3) it was also not predictable based on values of the random-effects predictor, and (4) because of this, the dependent variable was equally likely to take on either value, and was predicted to be equally likely to do so. Because the predictor varied within items, scrambling was done by simply randomly sorting the predictor values.

Nonetheless, similar results were obtained (Fig. 6.6): for a balanced sample, fixed-effects coefficients derived from both models are very similar (as a result, prediction accuracies across different scramblings of the fixed-effects predictor are highly correlated on the left in Fig. 6.6, $r = 0.88$), while for the corpus-like sample the mixed-effects model is better at predicting the test data with the real fixed-effects predictor than with the scrambled version (the mixed-effects model achieves better prediction accuracy with the real predictor 87% of the time, while the fixed-effects-only model does it only 21% of the time; the difference between the models is highly significant: $\chi^2(1) = 554.5$, $p < 0.00001$). The fixed-effects-only model frequently achieves equally low predictive accuracy with either predictor (62% of replications). Corpus-like sampling thus greatly diminishes the predictive power of the fixed-effects-only model (reducing number of times that the model with the correct predictor achieves better prediction accuracy from 99.5 to 21%). Corpus-like sampling does also hurt the mixed-effects model (99.5 vs. 87% correct predictor advantage, $\chi^2(1) = 5.2$, $p = 0.02$) but to a much lower extent.

There is a strong correlation in real predictor advantages in prediction accuracy between the two models when the sampling scheme is balanced but the correlation breaks down when sampling is corpus-like. We suspect that this is due to the fixed-effects-only model often basing its coefficient estimates to a large extent on the large proportion of data that come from the one or two highly frequent items in the sample, which sometimes works out (when those items are typical) but often does not, whereas the mixed-effects model is able to partial out the variance due to items.

In additional simulations, which we cannot report here due to space constraints, we have also verified that these results hold for a continuous predictor that is uncorrelated with frequency of an item and can vary within an item, for a binary predictor that cannot vary within an item and does not correlate with frequency, and for binary and continuous predictors that have effects whose slope varies across items. Thus we suggest that the mixed-effects model is highly preferred for unbalanced corpus-like samples whatever the predictor type. With balanced sampling, there is no advantage to incorporating a random effect of item even if there is such an effect when the aim is to generalize to new items. In contrast, with unbalanced, corpus-like sampling, random effects are essential for obtaining fixed-effects coefficient estimates that can be used to predict behavior on unseen items.

6 Limitations

A possible limitation of the present work is the exclusive use of leave-one-out cross-validation. We do not wish to commit to leaving only one of the items out at any one time. Leaving out, say, a tenth of all the items is a plausible alternative [66] that could be explored in future work. The bootstrap [67], which differs from cross-validation in that the data are sampled with replacement in constructing test and training sets, is another possibility that we have been reluctant to pursue for the purposes of these simulations as it would mean that the model would occasionally be tested on data it has been trained on. Finally, it is also possible to evaluate model predictiveness using AIC [50] and marginal R^2 [51] or other measures of fit as long as the random-effects predictors are left out of the evaluated model.

7 Conclusion

As argued above, corpus data present challenges in model evaluation: the data are unbalanced (at least across items, but often also across speakers because some speakers use the patterns of interest more than others). Frequent items often behave differently from rare items and are more likely to be idiosyncratic. These characteristics of corpus data require the use of mixed-effects models (particularly, fitting a random effect of item) and incorporating a fixed effect of frequency. A random effect of item helps improve estimates of coefficients associated with the fixed effects, making them more predictive. A fixed effect of frequency allows for the possibility of frequent words behaving differently from rare words (or, more generally, for the possibility of lexical diffusion; see Kapatsinski [47], for reasons why frequency effects can sometimes be non-monotonic as a change spreads through the lexicon)

While we argue for inclusion of a random effect of item in the analysis of corpus data, we argue that evaluating the fit of the model should be done by examining how much variance is captured by the fixed effects alone. We believe that evaluating the fit of a model that includes the random effects is likely to overestimate the fit. As a result, the researcher is in danger of prematurely terminating the search for additional predictors to code (if a final model is being evaluated), to accept a simpler model than is warranted by the data (if models are being compared and selected), or, more generally, to assign too much probability mass to an overly simple model.

In our first simulation, we have shown that evaluating full mixed-effects models containing random effects on fit to the training data does not allow one to select a model containing a real fixed-effects predictor over a model that contains a predictor whose values have been randomly scrambled. The problem is that when the fixed-effect predictors in the model are not predictive, the random effects can "step in" to capture the variance, allowing the model to still fit the data well. Nonetheless, this model would be useless in generalizing to unseen items that it was not trained on.

When we test a model on unseen items, the random effect of item may a priori appear to be of no use. Our simulations show that this intuition is incorrect. Even when models are tested on how well they predict behavior on unseen items, a mixed-effects model containing item identity as a predictor has higher predictive power than a fixed-effects-only model, under certain conditions. Namely, this advantage of mixed-effects models is seen for highly unbalanced samples typical of corpus data but not for balanced experimental designs, confirming that incorporating a random-effects predictor is particularly important when the number of observations across the values of that predictor is unbalanced. Without the random effect of item, the model may base its estimate of how the speaker will behave largely on high-frequency items. This is a particularly important issue with linguistic data, since high-frequency linguistic items are precisely the ones that are likely to be exceptional. Mixed-effects models provide a way to deal with this issue by partialling out the variance due to individual items, allowing for better generalization to unseen items.

Acknowledgments We would like to thank the two helpful anonymous reviewers of this article, as well as Daniel Johnson for bringing the Nakagawa and Schielzeth's [51] paper to our attention. We further thank the audience of the LSD 2012 for helpful comments and questions.

References

1. Labov W (1969) Contraction, deletion, and inherent variability of the English copula. Language 45:715–762
2. Sankoff D (1988) Sociolinguistics and syntactic variation. In: Newmeyer F (ed) Linguistics: the Cambridge survey, volume IV. Cambridge University Press, Cambridge, pp 140–161
3. Shih S, Grafmiller J, Futrell R, Bresnan J (2015) Rhythm's role in genitive construction choice in spoken English. In: Vogel R, van de Vijver R (eds) Rhythm in phonetics, grammar and cognition. Mouton de Gruyter, Berlin

4. Cedergren H, Sankoff D (1974) Variable rules: performance as a statistical reflection of competence. Language 50:333–355
5. Tagliamonte S, Baayen RH (2012) Models, forests and trees of York English: was/were variation as a case study for statistical practice. Lang Var Chang 24:135–178
6. Tagliamonte S (2006) Analyzing sociolinguistic variation. Cambridge University Press, Cambridge
7. Bresnan J, Cueni A, Nikitina T, Baayen RH (2007) Predicting the dative alternation. In: Bouma G, Kraemer I, Zwarts J (eds) Cognitive foundations of interpretation. Royal Netherlands Academy of Arts and Sciences, Amsterdam, pp 69–94
8. Drager K, Hay J (2012) Exploiting random intercepts: two case studies in sociophonetics. Lang Var Chang 24:59–78
9. Johnson DE (2014) Progress in regression: why sociolinguistic data calls for mixed-effects models. Ms. Lancaster University. http://danielezrajohnson.com/johnson_2014b.pdf
10. Johnson DE (2009) Getting off the GoldVarb standard: introducing Rbrul for mixed-effects variable rule analysis. Lang Ling Compass 3:359–383
11. Sapir E (1921) Language. Harcourt Brace, New York
12. Bybee J (2001) Phonology and language use. Cambridge University Press, Cambridge
13. Bybee J, Scheibman J (1999) The effect of usage on degrees of constituency: the reduction of don't in American English. Linguistics 37:575–596
14. Clark HH (1973) The language-as-fixed-effect fallacy: a critique of language statistics in psychological research. J Verbal Learn Verbal Behav 12:335–359
15. Coleman EB (1964) Generalizing to a language population. Psychol Rep 14:219–226
16. Bybee J (2003) Mechanisms of change in grammaticization: the role of frequency. In: Joseph B, Janda R (eds) Handbook of historical linguistics. Blackwell, Oxford, pp 602–623
17. Bybee J (2002) Word frequency and context of use in the lexical diffusion of phonetically conditioned sound change. Lang Var Chang 14:261–290
18. Gafos A, Kirov C (2009) A dynamical model of change in phonological representations: the case of lenition. In: Pellegrino F, Marsico E, Chitoran I, Coupé C (eds) Approaches to phonological complexity. Mouton de Gruyter, Berlin, pp 219–240
19. Kapatsinski V (2010) Rethinking rule reliability: why an exceptionless rule can fail. In: Proceedings from the annual meeting of the Chicago Linguistic Society, vol 44, no. 2, pp 277–291
20. Phillips BS (1984) Word frequency and the actuation of sound change. Language 60:320–342
21. Pierrehumbert J (2001) Exemplar dynamics: word frequency, lenition, and contrast. In: Bybee J, Hopper P (eds) Frequency and the emergence of linguistic structure. John Benjamins, Amsterdam
22. Schuchardt H (1885) Über die lautgesetze: gegen die junggrammatiker. Oppenheim, Berlin
23. Pitt MA, Johnson K, Hume E, Kiesling S, Raymond W (2005) The Buckeye corpus of conversational speech: labeling conventions and a test of transcriber reliability. Speech Comm 45:89–95
24. Barth D (2015) To have and to be: function word reduction in child speech, child-directed speech and inter-adult speech. Ph.D. Dissertation, University of Oregon
25. Barlow GM (2013) Individual differences and usage-based grammar. Int J Corpus Linguist 18:443–478
26. Baayen RH (2001) Word frequency distributions. Kluwer, Dordrecht
27. Zipf GK (1949) Human behaviour and the principle of least effort. Addison-Wesley, Reading
28. Oldfield RC, Wingfield A (1965) Response latencies in naming objects. Q J Exp Psychol 17(4):272–281
29. Barabási A-L, Albert R (1999) Emergence of scaling in random networks. Science 286:509–512
30. Merton RK (1968) The Matthew effect in science. Science 159:56–63
31. Simon HA (1955) On a class of skew distribution functions. Biometrika 42:425–440
32. Yule GU (1925) A mathematical theory of evolution based on the conclusions of Dr. J. C. Willis. Philos Trans R Soc B 213:21–87

33. Newman MEJ (2005) Power laws, Pareto distributions and Zipf's law. Contemp Phys 46(5):323–351
34. Hay JB, Pierrehumbert JB, Walker AJ, LaShell P (2015) Tracking word frequency effects through 130 years of sound change. Cognition 139:83–91
35. Pierrehumbert J (2002) Word-specific phonetics. In: Gussenhoven C, Warner N (eds) Laboratory phonology 7. Mouton de Gruyter, Berlin, pp 101–140
36. Briscoe T, Copestake A (1999) Lexical rules in constraint-based grammars. Comput Linguist 25:487–526
37. Goldberg AE (1995) Constructions. Chicago University Press, Chicago
38. Pinker S (1984) Language learnability and language development. Harvard University Press, Cambridge
39. Erker D, Guy GR (2012) The role of lexical frequency in syntactic variability: variable subject personal pronoun expression in Spanish. Language 88:526–557
40. Sankoff D, Labov W (1979) On the uses of variable rules. Lang Soc 8:189–222
41. Gerard J, Keller F, Palpanas T (2010) Corpus evidence for age effects on priming in child language. In: Proceedings of the 32nd annual meeting of the Cognitive Science Society, pp 218–223
42. Sonderegger M (2010) Testing for frequency and structural effects in an English stress shift. In: Proceedings of the Berkeley Linguistics Society, vol 36. pp. 411–425. https://dx.doi.org/10.3765/bls.v36i1.3927
43. Stefanowitsch A (2008) Negative entrenchment: a usage-based approach to negative evidence. Cogn Linguist 19:513–531
44. Ambridge B, Pine JM, Rowland CF, Chang F (2012) The roles of verb semantics, entrenchment, and morphophonology in the retreat from dative argument-structure overgeneralization errors. Language 88:45–81
45. Ambridge B, Pine JM, Rowland CF (2012) Semantics versus statistics in the retreat from locative overgeneralization errors. Cognition 123:260–279
46. Kapatsinski V (2010) What is it I am writing? Lexical frequency effects in spelling Russian prefixes: uncertainty and competition in an apparently regular system. Corpus Linguist Linguist Theory 6:157–215
47. Kapatsinski V (submitted) Sound change and hierarchical inference. What is being inferred? Effects of words, phones and frequency. Available at http://blogs.uoregon.edu/ublab/files/2014/11/SoundChange-21fo1iz.pdf
48. Raymond WD, Brown EL (2012) Are effects of word frequency effects of context of use? An analysis of initial fricative reduction in Spanish. In: Gries ST, Divjak D (eds) Frequency effects in language learning and processing. Mouton de Gruyter, Berlin, pp 35–52
49. Barth D, Kapatsinski V (2017) A multimodel inference approach to categorical variant choice: Construction, priming and frequency effects on the choice between full and contracted forms of am, are and is. Corpus Linguist Linguist Theory 13:203–260. https://doi.org/10.1515/cllt-2014-0022
50. Burnham KP, Anderson DR (2002) Model selection and multimodel inference: a practical information-theoretic approach. Springer, New York
51. Nakagawa S, Schielzeth H (2013) A general and simple method for obtaining R2 from generalized linear mixed-effects models. Methods Ecol Evol 4:133–142
52. Baayen RH (2008) Analyzing linguistic data: a practical introduction to statistics using R. Cambridge University Press, Cambridge
53. Keune K, Ernestus M, Van Hout R, Baayen RH (2005) Social, geographical, and register variation in Dutch: from written MOGELIJK to spoken MOK. Corpus Linguist Linguist Theory 1:183–223
54. Kothari A (2007) Accented pronouns and unusual antecedents: a corpus study. In: Proceedings of the 8th SIGDial workshop on discourse and dialogue. Association for Computational Linguistics, Antwerp, pp 150–157
55. Lohmann A (2011) Help vs. help to: a multifactorial, mixed-effects account of infinitive marker omission. Engl Lang Linguist 15:499–521

56. Theijssen D (2009) Variable selection in logistic regression: the British English dative alternation. In: Icard T, Muskens R (eds) Interfaces: explorations in logic, language and computation. Springer, Berlin, pp 87–101
57. Antić E (2012) Relative frequency effects in Russian morphology. In: Gries ST, Divjak D (eds) Frequency effects in language learning and processing. Mouton de Gruyter, Berlin, pp 83–108
58. Antić E (2010) The representation of morphemes in the Russian lexicon. Dissertation, UC Berkeley
59. Yao Y (2011) The effects of phonological neighborhoods on pronunciation variation in conversational speech. Ph.D. Dissertation, UC Berkeley
60. Pitt MA, Myung IJ (2002) When a good fit can be bad. Trends Cogn Sci 6:421–425
61. Bresnan J, Ford M (2010) Predicting syntax: processing dative constructions in American and Australian varieties of English. Language 86:168–213
62. Riordan B (2007) There's two ways to say it: modeling nonprestige there's. Corpus Linguist Linguist Theory 3:233–279
63. Theijssen D, ten Bosch L, Boves L, Cranen B, van Halteren H (2013) Choosing alternatives: using Bayesian networks and memory-based learning to study the dative alternation. Corpus Linguist Linguist Theory 9:227–262
64. Albright A, Hayes B (2003) Rules vs. analogy in English past tenses: a computational/experimental study. Cognition 90:119–161
65. Bates D, Maechler M, Dai B (2012) lme4: linear mixed-effects models using S4 classes. R package version 0.999999-0
66. Kohavi R (1995) A study of cross-validation and bootstrap for accuracy estimation and model selection. In: International joint conference on artificial intelligence (IJCAI), vol 2. Morgan Kaufmann, San Francisco, pp 1136–1143
67. Efron B, Tibshirani R (1993) An introduction to the bootstrap. Chapman & Hall, London

Chapter 7
(Non)metonymic Expressions for GOVERNMENT in Chinese: A Mixed-Effects Logistic Regression Analysis

Weiwei Zhang, Dirk Geeraerts, and Dirk Speelman

Abstract This paper focuses on the alternative choice between literal and metonymic expressions for the concept GOVERNMENT from an onomasiological point of view. With the help of mixed-effects logistic regression analyses, this study models the binary designations for GOVERNMENT with the data from a self-built corpus of texts from newspapers and online forums in Mainland Chinese and Taiwan Chinese. Mixed-effects models also provide a way of accommodating the random-effect factors such as the verbs in the data. The statistical results unveil that the choice of literal vs. metonymic designations is a result of the complex interplay of a number of conceptual, grammatical/discursive and lectal factors and no single decisive factor would determine people's onomasiological choice.

1 Introduction

Agreeing on the cognitive nature of metonymy, researchers in Cognitive Linguistics are striving for highlighting the crucial role of metonymy behind the semantic structure of language. Metonymy has been widely recognized to be instrumental in sense development and meaning structuring both synchronically and diachronically. In general, metonymy research in Cognitive linguistics is primarily being pursued from the semasiological perspective, i.e. focusing on how a specific word has acquired its lexical/grammatical metonymic sense and how a metonymy, as

---

W. Zhang (✉)
Institute of Linguistics, Shanghai International Studies University, Shanghai, China

Research Unit of Quantitative Lexicology and Variational Linguistics, University of Leuven, Leuven, Belgium
e-mail: weiwei.zhang@shisu.edu.cn

D. Geeraerts · D. Speelman
Faculty of Arts, Research Group QLVL, KU Leuven, Belgium

© Springer International Publishing AG 2018
D. Speelman et al. (eds.), *Mixed-Effects Regression Models in Linguistics*,
Quantitative Methods in the Humanities and Social Sciences,
https://doi.org/10.1007/978-3-319-69830-4_7

a mechanism of lexical innovation, triggers the compositional meaning of a compound or an idiom [1].

However, a purely semasiological study on metonymy is not sufficient to describe the function of metonymy. We should combine the semasiological perspective with an onomasiological one. The latter helps to discover the different conceptual or lexical "pathways" through which a concept or a group of concepts has developed by going back to the respective source concepts [1]. In order to reveal the reasons why a metonymic expression has been used, the onomasiological point of view has to be taken into account [2]. It is impossible for us to find out the pragmatic effects of metonymies, such as referring in the most economic and relevant way, causing euphemistic or humorous effects, and so on, or to identify the reasons why speakers choose the metonymic expressions, unless we examine a metonymic expression in contrast with its non-metonymic counterpart from an onomasiological perspective [2]. Therefore, the onomasiological choice of metonymic vs. literal naming for a given concept is worth investigating.

If we argue for an *onomasiological* reorientation of conceptual metonymy study, two questions which naturally arise are: what is people's lexical/categorical selection for a given concept (target), and what factors govern people's choice of a preferred designation or the alternative?. Answers to such questions are fundamentally linked to a contextualized, pragmatic interpretation of onomasiology [3], which focuses on an investigation of *use*, i.e. the actual choices made for a particular name as a designation of a particular concept in real language. If scholars are interested in the onomasiological choices for a particular referent in actual language usage, they invariably, and unsurprisingly, need to situate the issue in *multivariate data* for the possible variation. As Geeraerts points out, "a variational analysis is unavoidable to factor out lectal variation from the corpus data, but it is also a necessary and natural part of Cognitive Linguistics, to the extent that lectal variation underlies a specific form of linguistic meaning" [4]. In this paper, our primary interest lies in the lectal variation in the choice of literal vs. metonymic designation for the referent GOVERNMENT in two of the lectal varieties[1] of Chinese, Mainland Chinese (MC) and Taiwan Chinese (TC). Then, in order to study the combined effects of multiple factors and to model differences among internal and external factors, it is best to rely on *quantitative statistical techniques*, which have come into language variation and change studies in the form of different methods, such as generalized mixed-effects models (this volume), random forests [7], multidimensional scaling [8–10], etc.

The primary aim in this paper is to provide a systematic way of disclosing factors influencing the choice of designating one concept, i.e. GOVERNMENT, by either a metonymic (e.g. *Washington*) or a literal expression (e.g. *government*). There has been much interest in the past few years on the issue of PLACE NAME FOR GOVERNMENT metonymies, especially CAPITAL/COUNTRY NAME FOR GOVERNMENT (to name just a few [11–19]). So far the work that has been done

---

[1] A lectal variety refers to all types of language varieties or *lects*, such as regional dialects, sociolects, basilects, acrolects, idiolects, registers and styles [5, 6].

is mainly from a semasiological perspective. It is not immediately clear, however, whether PLACE NAME FOR GOVERNMENT takes place or not in a certain context if it does not combine the semasiological approach with the onomasiological one. This case study attempts to show that it is essential and necessary to extend studies on metonymy in an onomasiological and variationist Cognitive Linguistics direction, with an emphasis on the lectal variation in the usage of PLACE NAME FOR GOVERNMENT metonymy.

2 Methodology

2.1 Data Collection

2.1.1 Corpus Design

The data for this case study were taken from a self-built corpus which includes texts from newspapers and online forums in two of the lectal varieties of Chinese.[2] For Mainland Chinese, the *People's Daily* and the *Tianya Club* were chosen for the newspaper and the online forum, respectively. For Taiwan Chinese, the *United Daily News* was selected for the newspaper and the *PTT* for the online forum. The *People's Daily* is a daily newspaper published worldwide. For this study, we selected the Chinese-language edition published in Mainland. As the main official newspaper of the Communist Party, it generally provides direct information on the policies and viewpoints of the Party. The *United Daily News* is a daily newspaper produced by the United Daily News (UDN) group and published in Taiwan. In terms of political alignment, it is strongly Pan-Blue, i.e. pro-reunification, and conservative. Both newspapers are quality newspapers. The *Tianya Club* is an internet forum, based in Mainland China, with many sections, e.g. Worldlook, Entertainment, Fashion, and Travel. The *PTT*, short for "Professional Technology Temple", which is currently located at National Taiwan University, is the largest Bulletin Board System in Taiwan. It provides multi-topic online discussions including politics, social affairs and entertainment. Because no special event happened in the time period during which the texts were retrieved, it is unproblematic to not restrict the data resources to the exact same time period. What is of particular significance for our concerns is to design a relatively balanced corpus in terms of numbers of valid observations among the four resources.

Two points on the corpus design should be pointed out here. First, in practice, we have restricted ourselves to certain sections of the four data resources to make the data collection more efficient. We excluded sections like entertainment, sports, and

---

[2]Texts from the four resources were captured with the help of several Python scripts. The *People's Daily* and the *Tianya Club* can be accessed at http://paper.people.com.cn/rmrb and www.tianya.cn/bbs/. The *United Daily News* and the *PTT* can be accessed at http://udn.com/NEWS/mainpage.shtml and http://www.ptt.cc/index.bbs.html. We thank Tom Ruette for his help on the Python script for downloading texts from the *Tianya Club*.

Table 7.1 Data resources

	Mainland Chinese (MC)	Taiwan Chinese (TC)
Newspaper	People's Daily (PD) (Aug 17–22, 2010) (376,084 characters)	United Daily News (UDN) (Aug 17–22, 2010) (309,313 characters)
Online forum	Tianya Club (TYC) (Aug 17, 2010) (547,599 characters)	PTT.tw (PTT) (Aug 17–Sep 10, 2010) (25,493 characters)

health, in which the concept GOVERNMENT has very low frequency of occurrences. Second, for the online forum resources, only original posts were downloaded. Quotes from previous posts were not included. Reproductions from news material were also removed to keep the online forum language data less infused with news language.[3] Table 7.1 presents information on the data resources.

2.1.2 Potential Expressions for GOVERNMENT and Data Retrieval

The next step was to establish lists of potential expressions for GOVERNMENT and then to search for all of them in the self-built corpus. Before that, we have to briefly explain the GOVERNMENT concept in the Chinese context. Government in contemporary Chinese is conceptualized broadly as an administrative and executive organ of the state at both central and local levels, and it is composed of legislative, administrative and judicial institutions; narrowly, it is equal to the administrative branch [20, 21]. The basic assumption here is that although different states vary in terms of political systems, when Chinese people talk about the government of a state, they unconsciously project their basic conceptualization of government onto it. As long as the conceptualization of governments from different states in Chinese does not have consequential heterogeneity, it is feasible to conduct an onomasiological study.

As mentioned before, the main focus of this study is the alternative choices between literal and metonymic designations for GOVERNMENT. For literal designations, there are two expressions, 政府 *zheng-fu* "government" and 当局 *dang-ju* "authorities". For metonymic designations, in this case study we only looked at a specific metonymy, PLACE NAME FOR GOVERNMENT. Three main categories of place names are included here: country names, capital names and official residences of the state leader or government. We do not claim that we reach a complete list of expressions for GOVERNMENT. Two lists of potential expressions for GOVERNMENT were obtained, a literal one and a PLACE FOR GOVERNMENT

---

[3]For the *Tianya Club*, the first post of most thematic discussions is a copied news article which is then followed by original posts, so we excluded all first posts from the *Tianya Club* to build the online forum dataset for Mainland Chinese.

one extracted from the internet.[4] Examples (1) and (2) present portions of these lists. The numbers in brackets indicate the number of expressions in each category. In this study we also included the place name 大陆 *da-lu* "Mainland" as a country name in the list. This term literally refers to the geographical area under the jurisdiction of the People's Republic of China, and it can be used to refer to "the Mainland side (the PRC government)" in official contexts with reference to Taiwan in order to avoid cross-strait[5] conflict on the One-China policy.

> (1) literal expressions (N=2)
> 政府 *zheng-fu* "government", 当局 *dang-ju* "authorities"
> (2) place names
> a. country names (N=241)
> 中国 *zhong-guo* "China", 日本 *ri-ben* "Japan", 美国 *mei-guo* "the United States", etc.
> b. capital names (N=209)
> 北京 *bei-jing* "Beijing", 华盛顿 *hua-sheng-dun* "Washington", 巴黎 *ba-li* "Paris", etc.
> c. official residences (N=25)
> 中南海 *zhong-nan-hai* "Zhongnanhai", 白宫 *bai-gong* "White House", 唐宁街 *tang-ning-jie* "Downing Street", etc.

Observations including the expressions in the lists were retrieved by Python scripts from the corpus. Because the self-built corpus is not segmented or parsed, spurious hits were manually removed. All valid observations were presented with one sentence of context. For ambiguous cases, we could always trace back to the original texts for further context.

2.1.3 Meaning Identification in Contexts

After cleaning up the observations retrieved from the corpus, we come up with 12,652 valid observations. Then, a manual identification procedure was conducted to select observations in which the expression at hand has a meaning of GOVERNMENT. More precisely, in the present study, the concept is restricted to NONLOCAL GOVERNMENT (or CENTRAL GOVERNMENT), as governments exist at different levels. Next, we will explain the procedure for each category in turn: literal expressions, country names, capital names and official residences.

---

[4] The country name list was extracted from http://zh.wikipedia.org/wiki/国家列表. The capital name list was extracted from http://zh.wikipedia.org/wiki/各国首都列表. The official residence names are cases like 中南海 *Zhongnanhai* "the official residence of Chinese government leaders", 白宫 *bai-gong* "White House", 唐宁街 *tang-ning-jie* "Downing Street". Note that some countries or capitals have different linguistic expressions in the two language varieties, for instance, *Washington* has the Chinese equivalents 华盛顿 *hua sheng dun* in MC and 华府 *hua fu* in TC, and *New Zealand* has the Chinese equivalents 新西兰 *xin-xi-lan* in MC and 纽西兰 *niu-xi-lan* in TC. All possible linguistic variants are included in the list of place names for this study.

[5] When discussing affairs between Mainland China and Taiwan, the expression *cross-strait* is often used as a general term in reference to the Taiwan Strait.

Table 7.2 The semantic structure of literal expressions (政府 zhang-fu and 当局 dang-ju)

Meanings	N	%
Nonlocal government	557	45.73
Local government	504	41.38
Unspecific government	112	9.20
Other readings	45	3.69
Total	1218	100

Literal Expressions

In total, 1218 observations with literal expressions were extracted from the corpus. The semantic structure of literal expressions (i.e. 政府 zhang-fu "government" and 当局 dang-ju "authorities") is displayed in Table 7.2. Generally, four main meanings are identified for the literal expression category. The most common meaning of the two literal expressions is "nonlocal government" of an explicit political entity; see (3)a. The literal expressions may also refer to local governments at different levels, such as local governments on the provincial or municipal levels as in (3)b. Sometimes the contexts did not provide any clues to identify which specific political entity the government belonged to; see (3)c. The final case includes observations in which the two literal expressions have other readings like those of obsolete governments that no longer exist as in (3)d.

(3) a. "nonlocal government"
日本[政府]可能会再推出新的振兴措施, 以刺激疲弱的经济。(UDN-Aug18)
The Japanese [government] may reintroduce new revitalization measures to stimulate the ailing economy.
b. "local government"
玉树强震突如其来, 灾区各级党委[政府]紧急动员。(PD-Aug20)
The Yushu earthquake struck so suddenly; Party committees and [governments] at all levels in the stricken region conducted emergency mobilization.
c. "unspecific government"
你是[政府]领导人你不想你的国家好吗? (TYC-Aug17)
If you were the [government] leader, wouldn't you wish your country well?
d. other readings
明朝李闯王反是不能推翻大明[政府]的。(TYC-Aug17)
During the Ming Dynasty, Li Zicheng rebelled but he could not topple the Ming [government].

To make the concept comparable, this onomasiological study only selected the cases with the meaning of "nonlocal government". At the same time, we noted down any acts of government mentioned in the sentence for each "nonlocal government" observation for the purpose of getting a better idea of what the duty of a nonlocal government is, how it functions normally, etc. Such information can in turn serve as a reference for the semantic identification of the place name category. For example, one may clearly notice that one of the duties of a nonlocal government is to introduce new economic measures from (3)a.

Country Names, Capital Names and Official Residences

For the metonymy identification procedure of observations with place names, we used an adaptation of the MIP [22, 23] and the annotation schema of Markert and Nissim [15, 16]:

> 1. Read the entire sentence to establish a general understanding of the meaning.
> 2. a. For the place name in the text, establish its meaning in context, that is, how it applies to an entity, relation, or attribute in the situation evoked by the text.
> b. For each place name, determine its basic contemporary meaning, i.e. a locative/geographical meaning for all cases.
> c. Decide whether the contextual meaning of the place name contrasts with the basic meaning but has a contiguous relationship with it. Consult previous research on place names [14, 15, 24–26] for potential contiguous relationship candidates, such as a contiguous relationship between place and government/citizens/events, etc.
> 3. If yes, mark the place name as metonymic (Meto=yes). Then note its metonymic target, like GOVERNMENT (NONLOCAL GOVERNMENT, MUNICIPAL GOVERNMENT), EVENT (SUMMIT, SPORTS ACTIVITY, EXHIBITION ...), CITIZENS, etc.

In the following, we will introduce the identification procedure for each place name category and various issues encountered in practice.

Country names have 10,428 observations in the corpus. Table 7.3 presents the semantic structure of country names identified in this study. The most dominant meaning is "territory/geographical entity", which takes up 83.87%; see (4)a. The semasiological structure of country names is not easy to capture. The complication lies in the subtle difference between two concepts, i.e. GOVERNMENT and STATE. The former is a major agent of the state and exists to carry out the day-to-day business of the state, while the latter is a sovereign body, which is served by a continuous succession of different governments [27, 28]. In brief, SOVEREIGN STATE is a broader concept than GOVERNMENT, because the former includes several basic elements like permanent population, a government and the capacity to enter into relations with other sovereign states [29]. In many cases, a sovereign body can also undergo a process of personification, as "an international person", which makes it less distinguishable from the case of a "nonlocal government" meaning. For example, *Japan* in (4)b. should be interpreted as a sovereign state and by personification it has an attribution of human characteristics, i.e. having the emotion of fear. There is no explicit clue in the context for us to interpret the experiencer of the emotion as the Japanese government. The experiencer might be a mixture of Japan as an international person and Japanese citizens. The country name *United*

Table 7.3 The semantic structure of country names

Meanings	N	%
Territory/geographical entity	8746	83.87
State as an international person	1283	12.30
Nonlocal government	295	2.83
Other metonymic targets	104	0.10
Total	10,428	100

States in this example is simply a modifier of the noun "aircraft" specifying its attribute of "belonging to the United States"; therefore it stands for the sovereign state as a whole instead of the United States government. The country name *Korea* in (4)c is of interest for this study. It metonymically refers to "the nonlocal government of Korea", as one of the functions of governments is decision-making on taxes. The sense differentiation between "state as an international person" and "nonlocal government" sometimes is not straightforward. In those cases, the acts of government noted down in literal expression observations of "nonlocal government" (e.g. generally, decision-making on social, economic, political, and bureaucratic affairs, to make and enforce laws; more specifically, to control or censor the press, to conclude an agreement, to issue a declaration) could be used as references for the judgment between the two meanings. We adopted a principle of minimization of place name metonymy in this study. Therefore, for indeterminate cases of country names, we do not code them as cases of "nonlocal government" meaning unless the action/event expressed in the context has been confirmed as an act of government, as noted from the literal expression category. Without a doubt, country names can be metonymically used for other meanings (e.g. ORGANIZATION, NATIONAL TEAM, EVENTS, CITIZENS). For example, *Nicaragua, Ukraine, Canada* and *Mexico* in (4)d all metonymically refer to the national teams of these countries.

(4) a. "territory/geographical entity"
过去10年在[加拿大]也似的海豹死亡案例。(UDN-Aug18)
Seal deaths occurred in [Canada] in the past ten years as well.
b. "sovereign state as an international person"
奇怪啊, [日本]为什么不怕[美国]航母在日本海演习? (TYC-Aug17)
So strange! Why isn't [Japan] afraid of that the [United States] aircraft carrier is in the Sea of Japan for exercises?
c. "nonlocal government"
[韩国]在此背景下提出征收 "统一税", 很容易引起国民的反感。
(PD-Aug18)
[Korea] (The government of Korea) proposed the "Unitary Tax" in this case, which may create resentment among citizens.
d. other metonymic targets
中华队预赛四连胜, 分别击败[尼加拉瓜]、[乌克兰]、[加拿大]及[墨西哥]。(UDN-Aug23)
Chinese Taipei won four times successively in the preliminary competition, and it defeated [Nicaragua], [Ukraine], [Canada] and [Mexico] (the teams from the four countries).

Table 7.4 displays the semantic structure of capital names. Among 980 observations of capital names, 94.98% cases literally indicate a locative sense with a geographical feature; see (5)a. Around 2.65% cases involve metonymic processes for other targets instead of NONLOCAL GOVERNMENT, e.g. LOCAL GOVERNMENT, TEAM, EVENT, CITIZENS, as in (5)b. Only 25 cases of capital names metonymically express the meaning of "nonlocal government" of a particular state; see (5)c. In terms of types, only four capital names are found with the metonymic meaning of "nonlocal government": *Beijing* (N=19), *Taipei* (N=3), *Moscow* (N=2) and *Washington* (N=1).

Table 7.4 The semantic structure of capital names

Meanings	N	%
Geographical range	929	94.98
Other metonymic targets	26	2.65
Nonlocal government	25	2.55
Total	980	100

Table 7.5 The semantic structure of official residences

Meanings	N	%
Geographical range	23	88.46
Nonlocal government	3	11.54
Total	26	100

Table 7.6 Frequencies of observations of NONLOCAL GOVERNMENT in different resources

	Mainland Chinese (MC)	Taiwan Chinese (TC)
Newspaper	190	302
Online forum	208	180

(5) a. "geographical range"
还有一些台商,把库存货拿来[北京]卖。(UDN-Aug23)
Some Taiwanese businessman sold stock goods in [Beijing].
b. "other metonymic targets"
今年[北京]启动城乡结合部50个重点村城市化建设。(PD-Aug20)
[Beijing] (the municipal government of Beijing) starts the urbanization of 50 major villages at the outskirts of the city this year.
c. "nonlocal government"
[北京]如今核准了慈济在大陆成立总会。(UDN-Aug23)
Now [Beijing] (the Mainland Chinese government) approves the inauguration of the Tzu Chi Foundation on the Mainland.

The corpus has 26 valid observations of official residences; see Table 7.5. Most of them have the literal meaning of "a geographical range" as in (6)a. Three of them, *Zhongnanhai* (N=2) (6)b and *White House* (N=1), metonymically refer to the nonlocal governments of Mainland China and the United States respectively.

(6) a. "geographical range"
国务院总理温家宝17日在[中南海]紫光阁会见日本前首相。(PD-Aug18)
Premier Wen Jiabao met with the former Japanese Prime Minister in the Ziguang Pavilion of [Zhongnanhai] on August 17th.
b. "nonlocal government"
[中南海]牵挂着舟曲灾区。(PD-Aug21)
[Zhongnanhai] (The Mainland Chinese government) is concerned about Zhouqu disaster area.

After the meaning identification procedure for both literal expressions and place names, a total of 880 cases with the meaning of "nonlocal government" remained. The frequencies of observations with "nonlocal government" meaning in different resources are presented in Table 7.6.

2.2 The Variables

The 880 observations of NONLOCAL GOVERNMENT were then annotated for the following variables. First, we will discuss the response variable, and then we will present the predictors in the statistical model.

2.2.1 The Response Variable Meto

In the analysis of the data, we statistically modeled the designations of the concept GOVERNMENT. What concerns us here are the binary designations of the concept: literal expressions or PLACE FOR GOVERNMENT metonymy. The binary designation of the concept is called the response variable in the statistical model and given the label Meto. This response variable has two possible values, yes or no (metonymic expressions or literal expressions). It is encoded automatically according to the expression in the observation after the meaning identification procedure illustrated in Sect. 2.1. If the expression is from the literal expression list, the observation is coded with the value no; and with the value yes otherwise. In the dataset of 880 observations, we have 558 cases of Meto=no and 322 cases of Meto=yes. Obviously, this is a slightly biased distribution with a percentage of 63.41% (cases of Meto=no) versus a percentage of 36.59% (cases of Meto=yes).

2.2.2 The Predictors

The following predictors that represent conceptual, grammatical/discursive and lectal variables were included in the statistical model.

Conceptual Variables

The Variable Con_gp

The variable Con_gp stands for "concept groups". Although the general concept in the present study is NONLOCAL GOVERNMENT, governments of different countries vary greatly in terms of many facets. Therefore, a subdivision of the general concept is proposed in the study. The basic assumption is that people's choice of designations between literal vs. metonymic expressions may be determined to some extent by which country's government they are talking about. For the present study, we would like to test the effect of this conceptual factor on the usage of PLACE FOR GOVERNMENT metonymy.

In all, 44 political entities are identified and their governments are grouped into five subgroups: self (N=469), gpb (N=150), counterpart (N=105), Asia (N=103), and neutral (N=53). This variable was encoded manually. The Mainland Chinese government was assigned the value of self for observations from Mainland Chinese, and the value of `counterpart` for observations from Taiwan Chinese. The Taiwan government was accordingly assigned the value of `self` for observations from Taiwan Chinese and the value of `counterpart` for observations from Mainland Chinese. The value `gpb` stands for "global power brokers", which includes the governments of the permanent members of United Nations Security Council, including the governments of the US, the UK, France and Russia. Governments of those countries which Chinese people may have neutral or indifferent attitude towards were assigned the value `neutral`, e.g. the governments of Belgium, Canada, etc. Finally, `Asia` was assigned to the governments of those countries which are geographical neighbors of China, like the Japanese government, the Korean government, etc.

Value assignment for this variable involves a number of complications. For example, some governments may have multiple values: in principle, the Mainland Chinese government should belong to `self`, `counterpart` and `gpb` groups. To make things simpler we excluded it from the `gpb` group. In addition, we presumed that people from Mainland China and Taiwan might have similar emotional attitudes towards certain countries (e.g. Con_gp=neutral), which might bias the data. In fact, emotional attitudes towards certain countries may diverge to some extent between people from the two lectal varieties. The coding of this variable is just a simplification for the purpose of testing the influence of the conceptual factor, i.e. different governments, on the naming choice. Based on the previous studies of capital name metonymy [19, 30], we expect one's own government (with the value `self`), to which people may feel closer, to have the lowest frequency of metonymic designation; while governments of counterparts or of "global power brokers", which are further from the self, should favor metonymic designation [12].

The Variable `Topic`

The variable `Topic` stands for the topic of the text from which the observation was extracted. Five values were assigned for this variable: `worldwide` (N=423), `socialaffairs` (N=192), `crossstrait` (N=124), `domesticpolicy` (N=91), and `finance` (N=50).

Some observations were automatically assigned the value for this variable when the "section restriction" of the data resources indicated the topic of the text. For example, we included the "Finance/Business" sections of the newspapers and online forums. Observations from this section were automatically given the value `finance`. Many sections, however, do not specify the topics of the texts in the section. For instance, three sections from the *People's Daily*, i.e. 综合 *zong-he* "News Roundup", 视点 *shi-dian* "Opinions" and 要闻 *yao-wen* "Daily Selection" were downloaded. None of the section names indicate their specific topics. A more refined classification of their topics was conducted semi-automatically: an Instance Based Learning algorithm, implemented in Python, proposed a topic based on the title of the text, and then we verified the propositions that had low certainty.[6] If the observation was in the title (20 cases in total; see the variable Locus), the topic was manually identified.

The purpose of this variable is to test the influence of topic on the choice of literal vs. metonymic designations for a concept. The hypothesis is that there might be more PLACE FOR GOVERNMENT metonymy in the worldwide topic and in the crossstrait topic, as in bilateral or multilateral international affairs, people tend to use juxtaposed metonymies [19]: that is, more than one metonymy used in one sentence.

Grammatical/Discursive Variables

The Variable Syn

The variable Syn was coded for the syntactic position of the expression in each observation. It has two possible values: when the expression is in a non-subject position, the value of Syn is assigned nonsb; when the expression is in the subject position, the value sb is assigned. In the dataset, we have 193 cases of Syn=nonsb, and 687 cases of Syn=sb, see (7) and (8) respectively.

(7) Syn=nonsb
   a. 美国[政府]的推力被认为是主要原因。(PD-Aug21)
   The thrust force from the [government] of the United States is regarded as the main reason.
   b. 你倒是一连串帮[菲律宾]打圆场, 说人家有派人来解释了, 很罕见了够了。(PTT-Aug28)
   You, opposite to what we should do, are trying so hard to smooth things out for the [Philippines] (the government of the Philippines) by saying that they have already done what they need to do, which is rare, so we should be satisfied.

(8) Syn=sb
   a.[南韩]为了发展英语教育吸引外国投资, 在济州岛规划了一座"济州全球教育城"。
   (UDN-Aug30)
   To develop English education and attract foreign investors, [Korea] (the government of Korea) plans to build a "Jeju Global Education City" in Jeju City.
   b. 韩国[当局]在叫嚷"武力统一"之余, 干脆把射程1500公里的巡航导弹部署到了"三八"线边。
   (PTT-Sep28)
   After calling for "unifying by force", the Korean [authorities] have deployed their cruise missiles with 1500 km range to areas close to the 38th parallel.

---

[6]We have segmented the titles of all texts based on the Chinese Lexical Analysis System from the Institute of Computing Technology (ICTCLAS, http://ictclas.org/index.html). The topic classification was based on the title of each text. Instance Based Learning classifies unseen texts into the category of its most similar text in a manually annotated corpus. Similarity between two texts is measured by representing each text as a vector in a Euclidian space and taking the cosine of the angle between the two vectors. For the current task, a 3-Nearest Neighbor approach was used. A formal introduction to Instance Based Learning can be found in Chapter 8 of Mitchell [31]. We thank Tom Ruette for his help on the topic-identification programming script.

The hypothesis is that when the expression is in a subject position of a sentence, it has a higher possibility of choosing place for government metonymy than when it is in a non-subject position. Previous semasiological research has already shown that place name metonymies may occur more often in subject position [14, 19, 25].

The Variable Locus

This variable Locus simply encodes whether an observation is drawn from the main body or the title of the text. It was encoded automatically with two possible values, i.e. mainbody (N=860) and title (N=20). As stated in Papafragou [32], metonymy can be regarded as an economical means to express information during communication. Hence, due to word limits, titles are expected to have a higher density of metonymy than the main body of the text.

Lectal Variables

The Variable LangVar

The variable LangVar deals with the lectal effect on the usage of PLACE NAME FOR GOVERNMENT metonymy in the two lectal varieties. It was encoded automatically and has two possible values: MC (398 cases) and TC (482 cases). The purpose of this variable is to test whether MC and TC have any significant difference in the choice of literal vs. metonymic expressions for NONLOCAL GOVERNMENT; or in other words, to test whether or not the lectal factor plays a role in people's usage of place name metonymy. Based on the findings from an earlier semasiological study [19], we expect that there might be more PLACE FOR GOVERNMENT in Taiwan Chinese.

The Variable Style

The variable Style stands for the two stylistic resources, news and online forums. It has two possible values: 492 cases with the value of news and 388 cases with the value of forum. One could speculate that the naming choice is influenced by stylistic effects. We did not have any specific expectations with regard to this variable.

2.2.3 Summary of the Variables

So far, we have introduced all the variables coded for the statistical model. Table 7.7 provides an overview of predictor conditions and predicted effects of all variables.

Table 7.7 Overview of the predictions to be tested in the regression analysis

Predictor (fixed effect)	Predictor condition	Predicted effect
Conceptual variables	`Con_gp=self`	disfavors `Meto=yes`
	`Topic=worldwide`	favors `Meto=yes`
Grammatical/discursive variables	`Syn=sb`	favors `Meto=yes`
	`Locus=title`	favors `Meto=yes`
Lectal variables	`LangVar=TC`	favors `Meto=yes`
	`Style=news`	favors or disfavors `Meto=yes`

2.3 The Mixed-Effects Logistic Regression Model

The statistical model used in the present study is a mixed-effects logistic regression model, which is a statistical modeling with a binary response variable (such as in this case study `Meto=yes` or `Meto=no`) and with multiple explanatory variables containing both fixed effects and random effects, i.e. mixed effects [33–36]. The analyses were carried out based on the `lme4` package in R [37]. In this section, we will explain the mixed model in three parts: the random effect, the model selection and diagnostics, and the output of the regression model.

2.3.1 The Random Effect: Verb

The predictors introduced in Sect. 2.2 are all fixed effects, which are unknown constants that we try to estimate from the data [35]. In the present study, we also included a random variable in the statistical model. When the expression is in the subject position, we coded the verb or the predicate in the sentence as the random variable `Verb`. If the expression is not in the subject position, the variable `Verb` was assigned the value `irrelevant`; otherwise, we noted down the specific verb as the value, e.g. 是 *shi* "be", 宣布 *xuan-bu* "claim", 制定 *zhi-ding* "establish", 采取 *cai-qu* "adopt". In total, the value of `Verb` has 383 different levels. There might be strong verb-specific trends in the data: some verbs inherently favor `Meto=yes` and some verbs favor `Meto=no`. We could regard these verbs as sampled randomly from populations of verbs, and replicating the data collection may involve some other verbs. At the same time, we are not particularly interested in the individual probabilities of these specific verbs, which might well vary substantially, but in the population averages of verbs. To control this verb-specific trend or "noise", we treated `Verb` as the random variable in the mixed-effects logistic regression model, so that the model algorithm would adjust the intercept estimate for each verb depending on its influence in the data.

Table 7.8 Summary statistics of mixed models without and with interactions

Summary statistic	Mixed model without interactions	Mixed model with interactions
Number of observations	880 (of which 322 "metonymic" and 558 "literal")	
AIC[a]	920.4	903.1
C	0.865	0.871
Somer's D_{xy}	0.730	0.739

[a]AIC of an intercept only model is 1157.9

2.3.2 Model Selection and Diagnostics

A forward selection procedure resulted in a model with the predictors Con_gp, Section, Syn, Style, and Locus, and with the interaction Con_gp:Style. The model was confirmed by the procedures of both bootstrapping validation and cross-validation in order to avoid overfitting the data. The model also contains a random intercept for Verb. The R code for the model is as follows:

```
lmer(Meto ~ Con_gp + Section + Syn + Style + Locus +
  Con_gp:Style + (1|Predicate), data = mydata, family = binomial)
```

Before interpreting the detailed results, we would like to show some general information for two models: the mixed logistic regression model without interactions and the mixed logistic regression model with interactions (see Table 7.8). Two important indices should be mentioned here: the C index and the Somer's D_{xy}.[7] For both indices, a value close to 1 indicates that the model has good predictive ability. Therefore, the overall quality of the mixed model with interactions is more satisfactory. At the same time, the mixed model with interactions is verified as a more promising model with substantially higher predictive power than a fixed-effects only model with interactions, which has a C index of 0.819 and a Somer's D_{xy} of 0.638, or a random effect only model, which has a C index of 0.862 and Somer's D_{xy} of 0.725. In the following section we will discuss the detailed results of the mixed-effects logistic regression model with the interaction of Con_gp:Style.

2.3.3 The Regression Output

Table 7.9 shows the statistical output of the final model in the present study. Some important remarks of the logistic regression can be found in Speelman and Geeraerts

---

[7]The C index (or concordance index), ranging from 0.5 to 1, is used to measure the predictability of the logistic regression model. It is the "area under the ROC curve" to quantify the power of the model's predicted values to discriminate between positive and negative cases. A C index of 1 indicates perfect prediction; a C index of 0.5 indicates random prediction [38]. The Somer's D_{xy} provides a rank correlation between the predicted probability and the observed responses ranging from 0 to 1.

Table 7.9 Statistical output of the mixed effect logistic regression (general model)

	Estimate Std.	Error	z value	Pr(>|z|)
(Intercept)	-2.5007	0.4877	-5.128	2.93e-07
Con_gp=counterpart	3.7240	0.4525	8.231	< 2e-16
Con_gp=gpb	1.9408	0.3875	5.009	5.47e-07
Con_gp=neutral	0.7256	0.4422	1.641	0.100803
Con_gp=Asia	1.0167	0.3925	2.590	0.009600
Topic=finance	-2.3558	0.7043	-3.345	0.000823
Topic=crossstrait	-0.1967	0.2976	-0.661	0.508708
Topic=domesticpolicy	-1.6979	0.3668	-4.629	3.68e-06
Topic=socialaffair	-1.2266	0.2934	-4.180	2.91e-05
Syn=sb	1.0821	0.4244	2.550	0.010779
Style=forum	1.5651	0.2994	5.228	1.72e-07
Locus=title	1.4812	0.6357	2.330	0.019799
Con_gp=counterpart:Style=forum	-2.5740	0.5855	-4.397	1.10e-05
Con_gp=gpb:Style=forum	-1.4238	0.4872	-2.923	0.003470
Con_gp=neutral:Style=forum	0.3332	1.4168	0.235	0.814067
Con_gp=Asia:Style=forum	-0.2011	0.5583	-0.360	0.718643

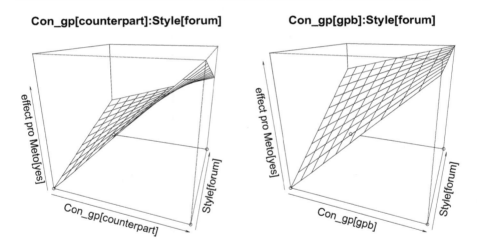

Fig. 7.1 Visualization of the interactions in the general model

[39]. The response of the model in this case is the logit of Meto=yes. The estimates are indeed on the logit scale and we used treatment contrast for the predictors. The reference levels of the predictors are Con_gp=self, Topic=worldwide, Syn=nonsb, Style=news, and Locus=mainbody.

Following the work of Speelman and Geeraerts [39], we graphically represent the interactions in the model as in Fig. 7.1. The x and y axes in the plots represent the interacting predictors and the z axis (the height) represents the joint effect of the two interacting predictors on the logit. On both x and y axes, the arrows run from the baseline values to the alternative values (the values in the brackets) of the predictors.[8] The significant interaction of Con_gp:Style indicates that a change of value in Con_gp systematically changes or reverses the effect of Style: compared to the baseline of Con_gp=self and Style=news the effect pro Meto=yes is slightly positive (logit=1.5651) in the case of Con_gp=self and Style=forum, more positive (logit=2.7151) in the case of Con_gp=counterpart and Style=forum, and most positive (logit=3.7240) in the case of Con_gp=counterpart and Style=news (see the left plot). Similarly, compared to the baseline of Con_gp=self and Style=news the effect pro Meto=yes is slightly positive (logit=1.5651) in the case of Con_gp=self and Style=forum, more positive (logit=1.9408) in the case of Con_gp=gpb and Style=news, and the most positive (logit=2.0821) in the case of Con_gp=gpb and Style=forum (see the right plot).

3 The General Regression Model for GOVERNMENT

In this section, we will interpret the statistical output of the mixed-effects logistic regression for the concept GOVERNMENT following the order of predictions listed in Table 7.7. Because a significant interaction was found with the predictors Con_gp and Style, these two predictors will be interpreted together (Sect. 3.2). First, however, we will comment on the general impact of the predictors in Sect. 3.1. Finally, the idiosyncrasy of the random variable will be discussed in Sect. 3.3.

3.1 General Impact of the Predictors

The relative importance of the predictors in this model is displayed in Fig. 7.2. The predictors are ordered according to their importance for their explained variation in the data. Apparently, Con_gp stands out as the most important predictor. The lectal factor Style as well as the conceptual factor Topic appear to have more impact on the variation than the grammatical factor Syn. The least influential predictor is the discursive factor Locus. The strong impact of the predictor Con_gp aligns with the expectation that governments of different sovereign states vary to a large extent with regards to speakers' choice of literal vs. metonymic expressions. This

---

[8]The 3D-graph visualization of the interaction was implemented in R [40]. Three more remarks need to be made about the z axis: "First, the plots are artificial in the sense that our predictors can assume only two possible values and that the only situations that can actually occur are represented by the four corners of the surfaces in the plot. Second, although in the plots the z axis is represented on a logit scale, we will describe the effects in terms of increased or decreased predicted probability of [Meto=yes]. Third, four small dots in the corners of each plot indicate the zero position on the z axis. This helps us to see whether joint effects are positive or negative" [29].

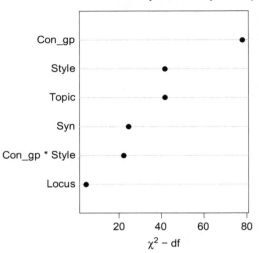

Fig. 7.2 Predictor importance in the model (fixed-effects only)

finding is quite intuitive. People do have different degrees of familiarity or emotional involvement with the governments of different states, and this conceptual difference may account for the difference in naming strategies. For example, for the states which people have no acquaintance with, their capital names or official residences of the governments may not be good candidates for metonymic expressions, as the hearer may not be able to process the metonymic link between the place name and the target.

3.2 Specific Influence of Fixed Effects

3.2.1 The Variable Con_gp and the Variable Style

For the interacting terms, the interpretation is less straightforward because the two predictors are no longer independent. With the help of the 3D graphs in Fig. 7.1, we find that: (1) (see both plots) when one's own government (Con_gp=self) is mentioned, it has the least probability of choosing metonymic expression in newspaper language; in other words, the probability of choosing metonymic expressions is much higher in online forums than in newspapers. (2) (see the left plot) whenever the observation mentions the counterpart government (Con_gp=counterpart), the probability of choosing metonymic expressions is lower in texts from online forums than from newspapers; the increase of probability of metonymic expressions when the predictor Con_gp runs from the baseline self to counterpart is less evident in the online forum sources than in the newspaper sources. (3) (see the right plot) whenever the observation mentions a "global power broker" government (Con_gp=gpb), the probability of choosing metonymic expressions is slightly higher in online forums than in newspapers; the probability of metonymic expressions increases less dramatically when the predictor Con_gp runs from the baseline self to gpb in online forums than in newspapers.

These findings confirm the predictions to some extent. Indeed, one is less likely to choose metonymic expressions for one's own government. However, it must be added that the way in which this happens is codetermined by the joint effect of the concept group and the stylistic status. We may expect more metonymic expressions even for one's own government when it is in an online forum situation. In other words, the probability differences between self and counterpart or between self and gpb governments diminish in the online forum language. Because the situation of Con_gp=self and Style=news has the least probability of choosing metonymic designations, we can speculate that the emotional involvement and the formality of language may jointly play a role here.[9] For different governments, people may have different degrees of emotional involvement or closeness, which in turn may contribute to the different designations for self and counterpart/gpb governments. The function of place name metonymy in creating or reflecting emotional distances has been discussed in the previous research: one's own capital, which is closest to the self, may serve as a metonymic source [12, 14, 19]. A conceptual metaphor might play a role in our thinking on this issue: CONCEPTUAL/EMOTIONAL DISTANCE (I.E. FAMILIARITY AND INVOLVEMENT) IS LINGUISTIC DISTANCE. For governments with which we have a close emotional attachment, we may tend to use the literal expression, which is more direct. For governments with which we have no emotional proximity, place name metonymy, as a less direct naming strategy, may provide a tool to reflect the distance. Newspapers, whose style is relatively formal, may also be influenced by certain political constraints and correctness, and journalists may have a closer stance towards their own government. Therefore, to show respect and care to their own government, when talking about it, it is perhaps no surprise then that journalists

---

[9]The interpretation of the relation between place name metonymy and emotional involvement towards specific governments is, of course, tentative. A careful and refined measurement of people's emotional attitudes is a must for a better appreciation of such relation. Apparently, the emotional involvement has both positive and negative sides. One may suspect that the two sides of emotional involvement could have quite different impacts on the choice of literal vs. metonymic expressions for GOVERNMENT. In the present study, we have not distinguished the specific effects of different kinds of emotional involvement, as it is very difficult to measure people's emotional attitudes toward the concept GOVERNMENT with the limited contexts. At the same time, individual journalists and online forum users may not have homogeneous types or degrees of emotional involvement towards governments. In addition, as Milić and Vidaković have proved, several reporter-related factors can influence the usage of CAPITAL FOR GOVERNMENT, for example, the reporter's whereabouts (abroad or home) and standpoint on the issue being reported [17]. One possible direction for further study on this issue would be a sentiment analysis of each text from which an observation is retrieved, which we would measure the positive, negative or neutral emotional attitudes of the journalist or online forum user toward the government in question, i.e. he/she is supporting or criticizing the government or stating a government-related affair in a neutral way.

tend to use the default, literal expression instead of a PLACE FOR GOVERNMENT metonymy. In online forums, however, language is less formal and users may feel less restricted in their language usage. At the same time, users may take a stance of emotional distance from their own government and they do not have to show their involvement in the relation with the government. Therefore, online forum users could also address their own and other governments in similar ways of by using place name metonymy. It is up to further research to pin down how these two factors codetermine people's usage of metonymy more exactly.

3.2.2 The Variable `Topic`

Compared with the baseline value `worldwide` (international affairs), the three topics of domestic policy, finance and social affairs noticeably disfavor metonymic designations for GOVERNMENT. In other words, in texts about international affairs, there is a relatively higher density of PLACE FOR GOVERNMENT metonymy. A possible interpretation for the variation found here is: multilateral affairs in an international situation involve multiple governments. In terms of number of characters, place name metonymy is shorter than the literal designation for GOVERNMENT in Chinese. For an economical purpose, we may refer to different governments metonymically by their country names or capital names, see (9), instead of the literal expressions in one sentence.

> (9) [俄罗斯]和[伊朗]当天还签署了关于组建合资企业联合管理核电站的文件，双方将各出一半资金。(PD-Aug22)
> [Russia] (The government of Russia) and [Iran] (the government of Iran) signed documents on setting up joint ventures for the joint management of nuclear power plants that day. They will both pay half of the funding.

3.2.3 The Variable `Syn`

The statistical model shows that the subject/non-subject distinction is also responsible for the naming choice for GOVERNMENT. The prediction that a metonymic designation is preferred in subject position is confirmed by the model. This provides supplementary evidence from an onomasiological perspective to the relation between metonymy and its syntactic position. Previous semasiological research has already shown that metonymies (esp. country, capital and organization names) seem more natural as subject [14, 19, 25, 41].

A tentative interpretation is provided for the positive correlation between metonymy and the subject position. The syntactic characteristic of metonymy may be attributable to the interaction of metonymy and agency. As we know, the deployment of metonymy (in this case, using place names) may generate ambiguity [42]. This ambiguity, however, may camouflage or blur the responsibility and accountability of the government by using place name metonymy to shift agency from the actor, i.e. the government ("a group of people"), to a non-agent locative entity, i.e. the country or the capital [43, 44]. Yamamoto studies the manipulation of the expression of agency as political maneuvers in media and points out that the attribution of agency means an assignment of responsibility [45]. Therefore, the impersonality of place names could reduce the accusation that the government is responsible for particular consequences.

A collateral consequence of this interpretation is that the relation between metonymy and subject position cannot be considered in isolation. The side effect of predicates has to been taken into account. Intrinsically, some predicates (e.g. 宣称 "claim", 宣布 "announce", 决定 "decide") require an agentive subject, while others (e.g. 是 "be", 需要 "need", 遭受 "suffer") do not have this requirement. It is legitimate to ask whether individual predicates in sentences affect the agentivity of the subject, which in turn affects our choice of literal vs. metonymic designations. This is one of the reasons why we have included Verb as a random variable in the mixed-effects model. We will comment on the random variable Verb in Sect. 3.3.

3.2.4 The Variable Locus

The estimate shows that compared to the main body of texts, the title favors metonymic designation. This finding upholds the effective and economical communicative function of metonymy. Many scholars have already pointed out the linguistic economy of the use of metonymy [46–48]. With word limits, the title may have a higher density of place name metonymy, which is shorter in number of characters than the literal expressions, see (10).

(10) 政治谈判 [北京]不急 (UDN, Aug-21)
  [Beijing] (The Chinese government) shows no anxiety about the political negotiations.

3.2.5 The Variable LangVar

Regrettably, predication that there is significant variation between Mainland Chinese and Taiwan Chinese is not borne out in this general model for GOVERNMENT.

3.3 The Random-Effect Variable of Verbs

The random variable in the model, Verb (in total, 383 levels), has a variance of 0.462 with a standard deviation of 0.680. Each level of the random variable has an adjustment to the intercept of the model. Table 7.10 presents the top five adjustments that increase the probability of Meto=yes, and those that encourage Meto=no. Each adjustment is added to the intercept estimate of the model to get the estimated value of Meto=yes for the observation with that verb.

Table 7.10 Top five pro Meto=yes and pro Meto=no adjustments for Verb as random effects

PLACE FOR GOVERNMENT metonymy (Meto=yes)	Literal expression (Meto=no)
签署 qian-shu "sign" 1.281	征收 zheng-shou "impose" −0.577
反对 fan-dui "be opposed to" 0.936	表示 biao-shi "express" −0.564
解决 jie-jue "solve" 0.779	管控 guan-kong "control" −0.491
举行 ju-xing "hold" 0.634	维持 wei-chi "maintain" −0.491
炒作 chao-zuo "sensationalize" 0.634	干预 gan-yu "intervene" −0.433

When we discussed the predictor Syn, we mentioned that the inherent properties of predicates may determine the agentivity of subjects, which in turn plays a role in the naming choice for GOVERNMENT in the subject position. The adjustments of random effects in the present model do not show a clear division of predicates between cases pro Meto=yes and cases pro Meto=no in terms of the requirement of agentivity on subjects. All the verbs listed in Table 7.10 require agentive and animate subjects. However, it is not always easy to measure the agentivity of the subjects required by a given verb in a quantitative way.

By including the random variable Verb as compensation for the intercept we assume that there is a random-effect factor of verbs. With a close examination of the specific activity involved, we may find that these verbs show idiosyncrasies: the verbs 签署 "sign", 反对 "be opposed to", and 解决 "solve" often involve topics of international affairs, which boost the usage of metonymic expressions, according to the fixed effect estimates. While, the verbs 征收 "impose (tax)", 管控 "control (the press)", 维持 "maintain", 干预 "intervene" frequently appear in texts with domestic policy and finance topics, which disfavor PLACE FOR GOVERNMENT metonymy (see discussion on the predictor Topic). The mixed-effects model keeps these verb-specific effects apart and precludes the random-effect factor of verbs from influencing in the estimates of fixed effects.

4 The Separate Regression Model for MAINLAND CHINESE GOVERNMENT

In Sect. 3, we interpreted the statistical output of the model of all 880 observations for the concept GOVERNMENT. As shown in the model, no significant lectal variation is found between the two language varieties of Chinese. Section 4 will discuss a separate model for a specific concept, MAINLAND CHINESE GOVERNMENT, in order to find whether hidden lectal variation exist between Mainland Chinese and Taiwan Chinese.

When we zoom in on the MAINLAND CHINESE GOVERNMENT (MCGOV), we come up with 308 valid observations. Among these 205 cases are from Mainland Chinese and 103 are from Taiwan Chinese. Figure 7.3 shows that the predictor LangVar seems to have an effect on the presence of metonymic designation

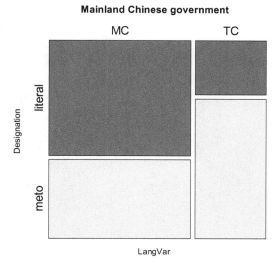

Fig. 7.3 The predictor LangVar affecting the choice of designation for MCGOV

Table 7.11 Statistical output of the mixed effect logistic regression (separate model)

	Estimate Std.	Error	z value	Pr(>|z|)
(Intercept)	-1.23352	0.55812	-2.210	0.027097
LangVar=TC	2.48711	0.44462	5.594	2.22e-08
Style=forum	0.38634	0.58083	0.665	0.505948
Syn=sb	0.04543	0.60886	0.075	0.940527
Locus=title	1.80832	1.21884	1.484	0.137904
LangVar=TC:Style=forum	-2.18609	0.58391	-3.744	0.000181
Style=forum:Syn=sb	1.28968	0.65553	1.967	0.049139

for this specific concept: Taiwan Chinese favors place name metonymies while Mainland Chinese favors literal ones. To confirm this hypothesis quantitatively, we fitted a mixed-effects logistic regression model with the same response variable and all the predictors except Con_gp from the general model (Con_gp brings the multicollinearity problem to the model).

4.1 The Separate Mixed-Effects Model

The output of the separate mixed-effects model with Verb as the random effect (166 levels) is displayed in Table 7.11. The R code for the separate mixed-effects model is:

```
lmer(Meto ~ LangVar + Style + Syn + Locus + LangVar:Style +
   Style:Syn + (1|Predicate), data = cndata, family = binomial)
```

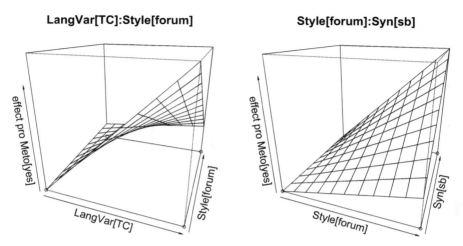

Fig. 7.4 Visualization of the interactions in the separate model

The summary statistics show that the model has better predictive power and explanatory strength (AIC=379.9, C=0.830, Somer's D_{xy}=0.661) than a fixed-effects only model, which has a C index of 0.741 and a Somer's D_{xy} of 0.483. The AIC of an intercept only model is 428.86.

In this separate model, the predictor Topic was not significant, and was eliminated from the model. LangVar has been included and shows a significant influence on the response variable. However, the interpretation of the predictor LangVar needs to be done with caution, as there is a strong interaction between LangVar and Style. In addition, Style also interacts with the predictor Syn. To provide a better understanding of the interactions in the separate model, we present them in Fig. 7.4, which gives a 3D graphical visualization of the joint effects. We will mainly interpret the joint effects of LangVar and Style in the next section.

4.2 The Lectal Variation Between Mainland and Taiwan Chinese

The significant interaction between LangVar and Style reveals this: compared to the baseline of LangVar=MC and Style=news the effect pro Meto=yes is slightly positive (logit=0.3863) in the case of LangVar=MC and Style=forum, more positive (logit=0.6874) in the case of LangVar=TC and Style=forum, and most positive (logit=2.4871) in the case of LangVar=TC and Style=news. In other words, the distinction between Mainland Chinese and Taiwan Chinese in terms of metonymic usage only has an strong effect in the context of newspaper texts. In online forum texts, this lectal effect is a minor one (the increase of probability of Meto=yes is unsubstantial when the value of LangVar runs from

MC to TC). The effect of online forum texts as a trigger for place name metonymy is restricted to Mainland Chinese. For Taiwan Chinese, the effect is reversed: the probability of place name metonymy decreases abruptly when the value of `Style` runs from `news` to `forum`.

In brief, the effect of the lectal factor, `LangVar`, is strongly dependent on the stylistic status of the text. `LangVar` contributes even more notably to the increase of metonymic designations in the newspaper language than in the online forum language. Put differently, we see a great divergence with regard to the choice between literal and metonymic designations for the MCGOV in the newspaper language between Mainland Chinese and Taiwan Chinese. For the online forum language, these two language varieties show less divergence on this issue, although the Taiwan online forum still has a slightly higher probability of containing metonymic designations than the Mainland online forum. We may speculate that the joint effect of `LangVar` and `Style` found here might not be arbitrary, but ideologically determined.

First, the concept MCGOV is the `self` government in Mainland Chinese, but the `counterpart` in Taiwan Chinese. Conceptually, people from the Mainland and Taiwan have different emotional involvement and conceptual closeness towards it. As presented in the general model, when it is about one's own government, people are more reluctant to choose a metonymic designation. More support for this explanation comes from the mixed-effects logistic regression model for observations with the concept ALL OTHER GOVERNMENT EXCEPT MAINLAND AND TAIWAN GOVERNMENTS (i.e. with the values `gpb`, `neutral` and `Asia` for the predictor `Con_gp`, in total 306 observations), where `LangVar` displayed no significant variation. Only for MCGOV does the lectal variation between Mainland Chinese and Taiwan Chinese emerge. One may state that people from the two language varieties may share certain similarities in their conceptualization of other governments; however, in respect of each others' governments, their conceptual closeness differs to a large extent.

Second, considerable research has explored the relationship between language and ideology, especially in the press [49–54]. The consensus is that there is a close relationship between language and ideology in media discourse [51]. The linguistic choices made in news articles may carry ideological meaning. In Mainland Chinese, the press (e.g. in this case, newspapers) has to follow more strict language regulations, which to some extent normalize the language usage in newspapers. To display language formality and to show their closeness toward MCGOV, journalists from Mainland may tend to use the literal expression for MCGOV. It is no wonder that place name metonymy has the least probability in Mainland newspapers.

A final remark on the effect of `LangVar` in the model is that, rather than providing clear evidence for lectal variation, the separate model offers a slightly different perspective on the variation, a perspective according to which one might speak of a kind of regional variation, but with some caveats. What we see in the model is not what we would traditionally call regional onomasiological variation. This is not 'formal onomasiological regional variation' where people from different regions would tend to choose another term for the same concept. Nor is it 'conceptual onomasiological regional variation' where the language users from different regions would tend to choose to construe the referent differently. In the present context, it simply is a given that people will construe MCGOV differently depending on whether live in the Mainland or Taiwan, because the role between that government and the language user is different by design. Therefore, one might argue that the lectal variation here is to some extent conceptual variation.

5 Summary

On the basis of a self-built corpus of newspapers and online forums from Mainland Chinese and Taiwan Chinese, we have performed a statistical analysis of factors contributing to the choice of literal vs. metonymic designations for the concept GOVERNMENT from an onomasiological perspective. With the help of multivariate analysis—in this case, a mixed-effects logistic regression model—this study unveils that the choice of literal vs. metonymic designations is a result of a complex interplay of a number of conceptual, grammatical/discursive and lectal factors and that no single decisive factor would determine people's onomasiological choice. Two separate models were built to detect the determining factors.

In the general model of all observations, we have shown that the choice of place name metonymy for GOVERNMENT is fostered by worldwide topics rather than topics of finance, domestic policy or social affairs, by expressions in subject position rather than non-subject position, and by a title context rather than a main body context. At the same time, the interaction between the concept group (i.e. different governments) and the stylistic status of the text shows that PLACE FOR GOVERNMENT metonymy is boosted dramatically when the observations mention the counterpart government in newspaper language. However, as the lectal effect of language variety might be absorbed by the predictor Con_gp, the general model cannot confirm the effect of LangVar, which prompted us to set up a separate model for observations of MCGOV to explore the lectal variation.

In the separate model of observations for MCGOV, significant variation between Mainland Chinese and Taiwan Chinese has been found and the language variety actually has the most impact on the statistical prediction. In addition, an important interaction between the language variety and the style of texts has to be added to the interpretation of lectal variation. Generally speaking, the lectal variation between Mainland Chinese and Taiwan Chinese is more evident in regard to place name metonymy in newspaper language rather than in online forum language. This lectal variation may link up with an ideological difference between people from the Mainland and Taiwan towards the concept under discussion, i.e. MCGOV. Apparently, Mainland people may have more conceptual closeness to MCGOV than people from Taiwan. This emotional distance could be reflected in people's linguistic usage, i.e. CONCEPTUAL/EMOTIONAL DISTANCE IS LINGUISTIC DISTANCE, which could be treated as one of the functions of metonymy. On the other hand, the strict language regulations in the press adopted in Mainland China may also contribute to the underused place name metonymy for MCGOV.

Establishing such an ideological interpretation of this lectal variation, of course, needs more evidence than what we have done in the present study. Although the *People's Daily* is a good representative of standard languages in Mainland newspapers, as there is no much divergence in political stances in the Mainland press, the political standpoints vary greatly in Taiwan newspapers, which may in turn influence people's usage of language. The data resource for Taiwan newspaper in the present study is the *United Daily News*, which has a pro-reunification standpoint toward Mainland China. To test whether ideological or political attitude would impact the use of place name metonymy for GOVERNMENT, the pro-independence newspaper *Liberty Times* from Taiwan should be included as another representative of the Taiwan press. If ideology indeed plays a role in the choice of literal vs. metonymic designations for GOVERNMENT, it is perfectly logical to assume that there is a significant difference between the *United Daily News* and the *Liberty Times* in their usage of place name metonymy for MCGOV.

But we have to keep in mind that the lectal variation found in the separate model has to be interpreted with caution. The concept that people are naming, i.e. MCGOV, is not the same thing to them. Therefore, it is at least debatable to call the observed variation regional. One might argue that the actual effect is not regional, but rather conceptual.

In sum, the present study shed some light on the current metonymy research in Cognitive Linguistics from the following two perspectives. First, theoretically it shows that the usage of metonymy is not an isolated linguistic or cognitive phenomenon but highly contextualized. In spite of its productivity, the (non-)application of PLACE NAME FOR GOVERNMENT metonymies is under conspicuous constraints, which are conceptually, discursively, grammatically and lectally motivated. Second, methodologically, this study responds to the call of Cognitive Linguistics to adopt a usage-based empirical methodology [55–57] by employing a corpus-based study and multivariate statistical analyses.

References

1. Blank A (2001) Words and concepts in time: towards diachronic cognitive onomasiology. metaphorik.de (01):6–25
2. Sweep J (2012) The onomasiological side of metonymy. In: Genis R et al (eds) Between west and east. Festschrift for Wim Honselaar. Uitgeverij Pegasus, Amsterdam, pp 611–631
3. Grondelaers S, Geeraerts D (2003) Towards a pragmatic model of cognitive onomasiology. In: Cuyckens H, Dirven R, Taylor JR (eds) Cognitive approaches to lexical semantics. Mouton de Gruyter, Berlin, pp 67–92
4. Geeraerts D (2005) Lectal variation and empirical data in cognitive linguistics. In: Ruiz de Mendoza Ibáñez FJ, Peña Cervel MS (eds) Cognitive linguistics: internal dynamics and interdisciplinary interaction. Walter de Gruyter, Berlin, pp 163–189

5. Geeraerts D (2006) Methodology in cognitive linguistics. In: Kristiansen G, Achard M, Dirven R, Ruiz de Mendoza Ibáñez FJ (eds) Cognitive linguistics: current applications and future perspectives. Mouton de Gruyter, Berlin, pp 21–49
6. Kristiansen G, Dirven R (2008) Cognitive sociolinguistics: language variation, cultural models, social systems. Mouton de Gruyter, Berlin
7. Tagliamontea SA, Baayen H (2012) Models, forests, and trees of York English: was/were variation as a case study for statistical practice. Lang Var Chang 24(02):135–178
8. Levshina N (2011) Doe wat ja niet laten kan: a usage-based Analysis of Dutch causative constructions. University of Leuven, Leuven
9. Ruette T (2012) Aggregating lexical variation: towards large-scale lexical lectometry. University of Leuven, Leuven
10. Zhang W, Geeraerts D, Speelman D (2015) Visualizing onomasiological change: diachronic variation in metonymic patterns for woman in Chinese. Cogn Linguist 26(2):289–330
11. Brdar-Szabó R, Brdar M (2011) What do metonymic chains reveal about the nature of metonymy? In: Benczes R, Barcelona A, Ruiz de Meñdoza FJ (eds) Defining metonymy in cognitive linguistics: towards a consensus view. John Benjamins, Amsterdam, pp 217–248
12. Brdar M (2006) Metonymic friends and foes, metaphor and cultural models. In: Benczes R, Csábi S (eds) The metaphors of sixty. Papers presented on the occasion of the 60th birthday of Zoltán Kövecses. Department of American Studies, School of English and American Studies, Eötvös Loránd University, Budapest, pp 75–83
13. Brdar M (2007) How to do a couple of things with metonymy. In: Cap P, Nijakowska J (eds) Current trends in pragmatics. Cambridge Scholars, Cambridge, pp 2–32
14. Brdar M, Brdar-Szabó R (2009) The (non-) metonymic use of place names in English, German, Hungarian, and Croatian. In: Panther K-U, Thornburg LL, Barcelona A (eds) Metonymy and metaphor in grammar. John Benjamins, Amsterdam, pp 229–257
15. Markert K, Nissim M (2003) Corpus-based metonymy analysis. Metaphor Symb 18(3):175–188
16. Markert K, Nissim M (2006) Metonymic proper names: a corpus-based account. In: Stefanowitsch A, Gries ST (eds) Corpus-based approaches to metaphor and metonymy. Mouton De Gruyter, Berlin, pp 152–174
17. Milić G, Vidaković D (2007) Referential metonymy of the type CAPITAL FOR GOVERNMENT in Croatian. In: Kosecki K (ed) Perspectives on metonymy. Peter Lang, Frankfurt am Main
18. Taylor J (2002) Category extension by metonymy and metaphor. In: Dirven R, Pörings R (eds) Metaphor and metonymy in comparison and contrast. Mouton de Gruyter, Berlin, pp 323–348
19. Zhang W, Speelman D, Geeraerts D (2011) Variation in the (non)metonymic capital names in Mainland Chinese and Taiwan Chinese. Metaphor Soc World 1(1):90–112
20. Qiu X, Zhang H, Wang Z (1986) Dictionary of politics (丘晓, 张宏生, 王正萍,《政治学辞典》). Sichuan People's Publishing House, Chengdu
21. Xie Q (2003) Contemporary Chinese government and politics (谢庆奎,《当代中国政府与政治》). Higher Education Press, Beijing
22. Pragglejaz Group (2007) MIP: a method for identifying metaphorically used words in discourse. Metaphor Symb 22(1):1–39
23. Steen G, Dorst AG, Herrmann JB, Kaal A, Krennmayr T, Pasma T (2010) A method for linguistic metaphor identification: from MIP to MIPVU. John Benjamins, Amsterdam
24. Halverson SL, Engene JO (2010) Domains and dimensions in metonymy: a corpus-based study of Schengen and Maastricht. Metaphor Symb 25(1):1–18
25. Markert K, Nissim M (2009) Data and models for metonymy resolution. Lang Resour Eval 43(2):123–138
26. Musson G, Tietze S (2004) Places and spaces: the role of metonymy in organizational talk. J Manag Stud 41(8):1301–1323
27. Flint C, Taylor P (2007) Political geography: world-economy, nation-state, and locality. Pearson Education, Edinburgh
28. Laski HJ (1935) The state in theory and practice. The Viking Press, New York

29. Shaw MN (2003) International law. Cambridge University Press, Cambridge
30. Brdar M (2009) Metonymy-induced polysemy and the role of suffixation in its resolution in some Slavic languages. Ann Rev Cogn Linguist 7(1):58–88
31. Mitchell T (1997) Machine learning. McGraw-Hill, New York
32. Papafragou A (1996) Figurative language and the semantics-pragmatics distinction. Lang Lit 5:179–193
33. Agresti A (2002) Categorical data analysis. John Wiley, New Jersey
34. Baayen H (2008) Analyzing linguistic data: a practical introduction to statistics using R. Cambridge University Press, Cambridge
35. Faraway J (2006) Extending the linear model with R: generalized linear, mixed effects and nonparametric regression models. Chapman and Hall/CRC Press, Boca Raton
36. Pinheiro J, Bates D (2000) Mixed-effects models in S and S-PLUS. Springer Verlag, Berlin
37. Bates D, Maechler M, Bolker B (2010) lme4: linear mixed-effects models using S4 classes. R package version 0.999375-35. http://lme4.r-forge.r-project.org/
38. Hosmer DW, Lemeshow S (2000) Applied logistic regression, vol 354. Wiley-Interscience, Hoboken
39. Speelman D, Geeraerts D (2009) Causes for causatives: the case of Dutch doen and laten. In: Sanders T, Sanders T, Sweetser E (eds) Causal categories in discourse and cognition. Mouton de Gruyter, Berlin, pp 173–204
40. R Development Core Team (2010) R: a language and environment for statistical computing. http://www.r-project.org/
41. Peirsman Y (2006) Quantitative approaches to metonymy. Quantitative investigations in theoretical linguistics. Osnabrück, Germany
42. Deignan A (2005) A corpus linguistic perspective on the relationship between metonymy and metaphor. Style 39(1):72
43. Moran MG (2005) Figures of speech as persuasive strategies in early commercial communication: the use of dominant figures in the Raleigh reports about Virginia in the 1580s. Tech Commun Q 14(2):183–196
44. Riad S, Vaara E (2011) Varieties of national metonymy in media accounts of international mergers and acquisitions. J Manag Stud 48(4):737–771
45. Yamamoto M (2006) Agency and impersonality: their linguistic and cultural manifestations. John Benjamins, Amsterdam
46. Blank A (1999) Co-presence and succession. In: Panther K-U, Radden G (eds) Metonymy in language and thought. John Benjamins, Amsterdam, pp 169–191
47. Ruiz de Mendoza Ibáñez FJ (2001) Metonymy and the grammar: motivation, constraints and interaction. Lang Commun 21(4):321–357
48. Warren B (1999) Aspects of referential metonymy. In: Panther K-U, Radden G (eds) Metonymy in language and thought. John Benjamins, Amsterdam, pp 121–135
49. Berthele R (2008) A nation is a territory with one culture and one language: the role of metaphorical folk models in language policy debates. In: Kristiansen G, Dirven R (eds) Cognitive sociolinguistics: language variation, cultural models, social systems. Mouton de Gruyter, Berlin, pp 301–332
50. Fowler R (1991) Language in the news: discourse and ideology in the press. Routledge, London
51. Kuo S-H, Nakamura M (2005) Translation or transformation? A case study of language and ideology in the Taiwanese press. Discourse Soc 16(3):393–417
52. Dijk V, Teun A (1998) Ideology: a multidisciplinary approach. Sage, London
53. White M, Herrera H (2003) Metaphor and ideology in the press coverage of telecom corporate consolidations. In: Dirven R, Frank RM, Pütz M (eds) Cognitive models in language and thought: ideology, metaphors and meanings. Mouton de Gruyter, Berlin, pp 277–323
54. Wolf H-G, Polzenhagen F (2003) Conceptual metaphor as ideological stylistic means: an exemplary analysis. In: Dirven R, Frank RM, Pütz M (eds) Cognitive models in language and thought: ideology, metaphors and meanings. Mouton de Gruyter, Berlin, pp 247–275

55. Glynn D (2010) Quantitative methods in cognitive semantics: corpus-driven approaches, 1st edn. Walter de Gruyter, Berlin
56. Grondelaers S, Geeraerts D, Speelman D (2007) A case for a cognitive corpus linguistics. In: Gonzalez-Marquez M, Mittelberg I, Coulson S, Spivey MJ (eds) Methods in cognitive linguistics. John Benjamins, Amsterdam, pp 149–169
57. Tummers J, Heylen K, Geeraerts D (2005) Usage-based approaches in cognitive linguistics: a technical state of the art. Corpus Linguist Linguist Theory 1(2):225–261

CPSIA information can be obtained
at www.ICGtesting.com
Printed in the USA
LVHW03*2046100718
583290LV00009BA/222/P